The Psychology of Mathematics

This book offers an innovative introduction to the psychological basis of mathematics and the nature of mathematical thinking and learning, using an approach that empowers students by fostering their own construction of mathematical structures.

Through accessible and engaging writing, award-winning mathematician and educator Anderson Norton reframes mathematics as something that exists first in the minds of *students*, rather than something that exists first in a textbook. By exploring the psychological basis for mathematics at every level—including geometry, algebra, calculus, complex analysis, and more—Norton unlocks students' personal power to construct mathematical objects based on their own mental activity and illustrates the power of mathematics in organizing the world as we know it.

Including reflections and activities designed to inspire awareness of the mental actions and processes coordinated in practicing mathematics, the book is geared toward current and future secondary and elementary mathematics teachers who will empower the next generation of mathematicians and STEM majors. Those interested in the history and philosophy that underpins mathematics will also benefit from this book, as well as those informed and curious minds attentive to the human experience more generally.

Anderson Norton is Professor in the Department of Mathematics at Virginia Tech, USA, where he has been teaching mathematics courses for future teachers for 15 years. He is the editor of *Constructing Number* and the coauthor of *Developing Fractions Knowledge* and *Numeracy for All Learners*.

W0113238

The Psychology of Mathematics

A Journey of Personal Mathematical Empowerment for Educators and Curious Minds

Anderson Norton

Routledge
Taylor & Francis Group

NEW YORK AND LONDON

Cover image: © Eve Azano

First published 2022
by Routledge
605 Third Avenue, New York, NY 10158

and by Routledge
2 Park Square, Milton Park, Abingdon, Oxon, OX14 4RN

Routledge is an imprint of the Taylor & Francis Group, an informa business

Library of Congress Cataloging-in-Publication Data
Names: Norton, Anderson, author.
Title: The psychology of mathematics : a journey of personal mathematical empowerment for educators and curious minds / Anderson Norton.
Description: New York, NY : Routledge, 2022. | Includes bibliographical references and index. | Identifiers: LCCN 2021045516 | ISBN 9781032020716 (hardback) | ISBN 9781032020693 (paperback) | ISBN 9781003181729 (ebook)
Subjects: LCSH: Mathematics—Philosophy. | Mathematics—Psychological aspects. | Mathematics—Study and teaching. | Knowledge, Theory of.
Classification: LCC QA8.4 .N67 2022 | DDC 510.1/9—dc23/eng/20211123
LC record available at https://lccn.loc.gov/2021045516

ISBN: 978-1-032-02071-6 (hbk)
ISBN: 978-1-032-02069-3 (pbk)
ISBN: 978-1-003-18172-9 (ebk)

DOI: 10.4324/9781003181729

Typeset in Bembo
by Apex CoVantage, LLC

To Caroline and Eleanor, who have indulged my work as an integral aspect of our father–daughter relationships, that they might learn as much from me as I have learned from them.

Contents

Figures

Tables

About the Author

Anderson Norton is Professor of Mathematics Education in the Department of Mathematics at Virginia Tech. He was introduced to the work of Swiss psychologist Jean Piaget while completing his graduate studies under the direction of Les Steffe at the University of Georgia. Dr. Norton's research has become heavily influenced by Piaget's epistemology of mathematics, which characterizes mathematics as the coordination of mental actions that are reversible and composable. Dr. Norton builds psychological models of students' mathematics by identifying the mental actions students have available and how they coordinate those actions in solving mathematical tasks and developing mathematical concepts. He has authored more than 100 related publications, and his models have informed the design of curricular materials and instructional tools. With support from the National Science Foundation and the US Math Recovery Council, Norton participated in an interdisciplinary team that designed apps to support students' construction of fractions concepts. For such efforts to turn research into practice, he was awarded the Early Career Award by the Association of Mathematics Teacher Educators, in 2013. More recently, Dr. Norton has focused on fostering collaborations with psychologists and neuroscientists so that cognitive constructs, such as working memory, might inform mathematics education, and so that research in mathematics education might inform psychological studies of mathematical development. With psychologist Martha Alibali at the University of Wisconsin, he coedited the Springer book, *Constructing Number: Merging Perspectives from Psychology and Mathematics Education*. Norton has served as the chair of the steering committee for the *North American Chapter of the Psychology of Mathematics Education* and the chair of the editorial panel for the *Journal for Research in Mathematics Education*.

Acknowledgments

This book would not have happened without encouragement and critical feedback from colleagues and friends over many years, especially the following: Amy Azano, Brad Bassler, Robert Berry, Steven Boyce, Bud Brown, Anthony Cate, Chris Carmody, Chris Drumm, Amy Hackenberg, Peter Haskell, Nicola Hodkowski, Mandy Jansen, Signe Kastberg, Sarah Kerrigan, Keith Leatham, LouAnn Lovin, Kevin Moore, Nathan Phillips, Kyeong Hah Roh, Wendy Sanchez, Ted Shifrin, Marty Simon, Irma Stevens, Les Steffe, Ron Tzur, Catherine Ulrich, Megan Wawro, Michael Weiss, and Jesse Wilkins.

Introduction

Mathematical Empowerment

French poet Paul Valéry called mathematics "the science of acts without things—and through this, of things that one can define by acts."[1] This perspective stands in stark contrast to the commonly held notion that mathematical objects exist in their crystallized forms in some perfect world apart from us . . . or perhaps in a textbook. Valéry was chair of the French Academy, previously occupied by Anatole France, who critiqued the notion with the following anecdote: "Before there were feet and before there were posteriors in this world the kick in the posterior must have had existence for all eternity in the bosom of God."[2]

In rejecting this view and framing mathematics as a human activity, we confront a few serious philosophical questions. What is mathematics? Why does mathematical knowledge seem infallible and true, even when compared to scientific knowledge? For example, theories of gravity evolve, but the side lengths in a right triangle will always satisfy the equation $a^2 + b^2 = c^2$, absolutely and without exception. How do we, as humans, have access to such certain knowledge? These philosophical questions have psychological answers.

For millennia, mathematicians and philosophers have wondered at the power of mathematics. Historically, they have found their answers in the heavens. As Galileo put it,

> [p]hilosophy is written in this grand book—I mean the universe—which
> stands continually open to our gaze, but it cannot be understood unless one

first learns to comprehend the language in which it is written. It is written in the language of mathematics, and its characters are triangles, circles, and other geometric figures, without which it is humanly impossible to understand a single word of it; without these, one is wandering about in a dark labyrinth.

(Galileo, 1623, *The Assayer*)

In claiming that mathematics is written into the fabric of the universe, Galileo was following Plato—the ancient Greek philosopher who founded Platonism.[3] Plato espoused the belief that earthly knowledge amounts to a mere shadow of heavenly knowledge, where numbers and geometric objects reside. Although we might draw square figures, Plato believed that the true square exists only in the heavens—in the fabric of the universe. This view reduces mathematics to a language that mathematicians can pass on to their students, if only they would listen.

Math is a human creation. All of us have access to it because each of us can create it by reflecting on our own mental activity. We find the genesis of mathematics in infancy, and we find it developing in childhood, as children begin to discern shape and number. We see math build on itself as we construct ever more complex mathematical objects from simpler ones. We see the universe unfold, not because we are discovering its language but because we are constructing ever more powerful tools for organizing our experiences within it.

This book is about the nature of mathematics—not just the way we learn math but mathematics itself. In reading this book, you will experience how mathematical objects (e.g., numbers and shapes) arise from your own activity. These objects do not exist out there in the world. Rather, you create them by coordinating our own mental actions. You can then project them into the world in order to make sense of what you observe. Like finding constellations in the night sky, we find mathematical objects in the world because we impose them on our experiences in the world, in order to organize and structure those experiences. As we become aware of these projections, Platonism gives way to the power of our own constructions.

The book should be especially relevant to future mathematics teachers who will empower the next generation of mathematicians, STEM (science, technology, engineering, and mathematics) majors, and informed citizens. We empower our students by fostering their constructions of mathematical structures. To do that, we must first recognize the power students bring into our classrooms via their mental activity. Thus, this book focuses on a psychology of mathematics that reframes mathematics as something that exists first in the minds[4] of our students rather than something that exists first in a textbook.

Reflection

Take a look at Figure 0.1. What do you see?

What you see is what you've got. Young children might see a triangle of numbers with lots of 1s on the edges. They might also notice a column of numbers from 1 to 7 or a kind of symmetry across each row (e.g., 1, 4, 6, 4, 1). As they develop new ways of structuring numbers, they see more: multiples of 7 in the bottom row, triangular numbers in the third column (1, 3, 6, 10, 15, 21), and rows that sum to powers of two (e.g., $1 + 4 + 6 + 4 + 1 = 2^4$). These additional structures emerge, not from the triangle of numbers but from children's minds as they make sense of what they see.

The numbers in Figure 0.1 come from the first eight rows of Pascal's triangle. Aside from the 1s, each entry of the triangle is generated by adding the term directly above with the term above and to the left (e.g., the 20 comes from the two 10s). To be sure, this simple rule generates lots of interesting patterns that we might recognize. However, we can only recognize patterns if we have constructed mathematical structures for organizing those patterns. In other words, we have to impose our existing mental structures onto Pascal's triangle in order to see the patterns; the patterns do not impose themselves on us.

Reflection

Consider the array of numbers shown in Figure 0.2. Now what do you see?

As you might have guessed, the numbers in Figure 0.2 were randomly generated. There is no underlying rule for generating them. Yet, we still try to impose our structures on them. For example, you might have noticed a triangle of 2s in

1							
1	1						
1	2	1					
1	3	3	1				
1	4	6	4	1			
1	5	10	10	5	1		
1	6	15	20	15	6	1	
1	7	21	35	35	21	7	1

Figure 0.1 Finding patterns among a triangle of numbers.

4	1	4	3	9	3	7
6	2	8	7	7	4	5
9	6	1	2	2	7	3
5	9	8	8	2	4	6
6	4	5	5	1	5	5
4	6	7	3	4	2	4
9	3	2	6	5	6	9
1	8	5	8	4	3	5

Figure 0.2 Finding patterns among a rectangle of numbers

the middle or a lot of odd numbers in the right column. The patterns we try to impose on the figure might not hold up absolutely, but that does not stop us from trying. When our available structures fail to account for what we see, we conclude that either (a) the numbers are random or (b) we don't have enough mathematical knowledge to see the pattern. Both conclusions demonstrate reason, but the second one also demonstrates the humility of a mathematician.[5]

All apparent arrogance aside, mathematicians are, above all, humbled by the sense that their mathematical knowledge is never complete and maybe not even useful. Hardy captured the sentiment in the closing of his mathematical apology: "I have just one chance of escaping a verdict of complete triviality, that I may be judged to have created something worth creating."[6] For this reason, mathematicians must continually grow, to construct ever more powerful structures, on the basis of their own mental actions.[7] Mathematicians will even study the distribution of random numbers to determine the probability that they are, in fact, randomly generated.[8]

This book describes the psychological basis for mathematics at every level. We first learn to count through activities of pointing and reciting number words. As we grow, we learn to coordinate these actions. Further mathematical development builds on such early development through the coordination of new actions. Ultimately, even the most advanced mathematics is grounded in our own actions, which we learn to coordinate.

This book will simultaneously pursue two goals: (1) to elucidate your personal power in constructing mathematical objects based on your own mental activity and (2) to illustrate the power of mathematics in organizing the world as we know it. The book is intended not only for mathematics students and teachers but also for a broader audience of curious minds—anyone who has wondered about the apparent infallibility of mathematics or questioned their own ability to attain it.

The first few chapters of the book treat number and geometry separately. In later chapters, topics of geometry and number merge into algebra. This merger of shape and number follows both the psychological and historical trajectory of algebra. Still later chapters delve into more advanced mathematical topics, such as calculus and complex analysis. We will see that these ideas, too, have their basis in the coordination of our own mental actions.

From this perspective, equity is not an additional consideration but an integral one.[9] The position statement on access and equity from the National Council of Teachers of Mathematics (NCTM) begins as follows: "Creating, supporting, and sustaining a culture of access and equity require being responsive to students' backgrounds, experiences, cultural perspectives, traditions, and knowledge when designing and implementing a mathematics program and assessing its effectiveness."[10] Equity in mathematics demands that each of us has access and opportunity to realize our mathematical power and develop our personal identities as mathematicians.[11] Developing this personal power and mathematical identity is the entire focus of the book.

Along the way, we will find inequities in the history of mathematics, especially in the Western appropriation of mathematical ideas.[1] "Pascal's triangle" is one of many Western appropriations. The Indian mathematician (and poet), Acharya Pingala, studied this triangle, along with the binomial theorem around 200 BC, so we might better refer to it as Pingala's triangle.[12] Likewise, the Pythagorean theorem was known within ancient Egyptian and Babylonian cultures long before Pythagoras roamed Greece, and Gaussian elimination—named for the 19th-century German mathematician Carl Gauss—originated in ancient China centuries before Germany even existed. Recognizing such appropriations does not discount the substantial contributions of Western mathematicians, like Gauss, but rather, it acknowledges that when Newton stood on the shoulders of giants,[13] these giants included unnamed and forgotten mathematicians from Northern Africa, East Asia, and around the globe.

It is no coincidence that we find mathematical developments in centers of economic power like ancient Egypt and Babylon, and like present-day Europe, the United States, and China. Chinese mathematicians, in particular, have reemerged in prominence just as their nation has reemerged as an economic power. Equity and power cannot be disentangled from each other or mathematics. Here, we frame mathematics as a personal power that students construct when given the opportunity. Although opportunities vary across time and place, mathematics comes from them.

Chapter 1 elaborates on a simple example of how we construct mathematical objects, namely, numbers. Like all other mathematical objects, 7 is not out there in the world for us to see; we make 7 and use it to organize what we perceive as, say, seven yellow bricks. This chapter also explains some of the innate tendencies humans have, from birth, that support the construction of numbers. Chapter 4 returns to the construction of numbers. Using examples of fractions and decimals, the chapter demonstrates how we extend our concept of numbers by including new kinds of units and actions.

Chapter 2 illustrates how we can begin to construct mathematics on the basis of a single mental action: reflection. By reflecting on reflections, we can learn to coordinate pairs of those mental actions to produce rotations and translations as transformations of the plane. Chapter 3 extends this idea and connects it to the Erlangen program, which closely relates to the theme of this book. First proposed by mathematician Felix Klein in the late 1800s, the Erlangen program was used to classify different kinds of geometry. This chapter provides an illustration of that program with quadrilaterals and extends it as a psychological foundation for all of mathematics.

Chapter 5 demonstrates the psychological basis for one of the most famous theorems in mathematics: the Pythagorean theorem. The ancient Greek mathematician Euclid proved the Pythagorean theorem through a sequence of 47 propositions, but as the chapter demonstrates, these propositions—and the axioms from which they start—are grounded in action. On one hand, formalization and axiomatization mask the dynamic nature of mathematics. On the other hand, they provide for a rigorous (rule-based) method of communicating arguments. Likewise, algebra—the topic of Chapter 6—provides a rigorous method for symbolizing mathematical objects while offloading their demands on working memory. We can manipulate those symbols as a proxy for acting on mathematical objects, including unknown quantities.

Chapter 7 investigates the mental actions undergirding angles and demonstrates how we can coordinate those actions to prove a theorem about Pascal's mystic hexagon. Chapter 8 demonstrates how geometry and number become increasingly integrated as we extend the natural numbers and fractions to integers and real numbers. Geometric transformations introduced in Chapter 2 are used to construct and transform new kinds of numbers—numbers with directions. Ultimately, this leads to the treatment of numbers as vectors in a plane, including the plane of complex numbers.

Chapters 9, 10, and 11 focus on functions and relations between variables. The challenge is to understand how variables covary, as represented by an equation or a graph. Even as variables vary, some things remain the same, and this is the essence of the function or relation. Chapter 9 introduces a framework for understanding variables and covariation as a coordination of our own mental actions. Chapter 10 focuses on covariation in the context of trigonometric functions, concluding with a surprising connection to prime numbers, constructible polygons, and Pingala's triangle. Chapter 11 extends the idea of covariation to rates of change—a central idea in calculus. When Newton thought about and measured rates of change, his methods were much more intuitive than the methods students are taught in school today.

The book closes (in Chapter 12) with a return to the philosophical questions raised at the beginning of this introduction. In answering those questions, the chapter turns Platonism on its head. If we understand mathematics as a product of psychology, then how do we explain the uncanny power of mathematics in understanding the universe? For example, we might be able to explain how ellipses arise from the coordination of certain mental actions, but then why do ellipses fit the paths of planets? Mathematics does not come from the heavens but, rather, is something that we project out into them. We should then wonder how those projections fit so well.

Each chapter includes reflections—like the two reflections included in this introduction—and concludes with a set of activities. The reflections mark places for you to pause and think about your own mental actions. The activities are intended to further raise your awareness of the mental actions you coordinate when doing mathematics. For readers interested in further details and related research, footnotes are included within each chapter.

In sum, this book is about your mathematical power: a power to unlock ever more complex structures in an otherwise unknowable universe. Chapter by chapter, you should experience new mathematical power or at least become more aware of the mathematical power you already have.

NOTES

1. This quote comes from Valéry's (1935) in *Cahiers, Volume II*, as cited and translated by Ernst von Glasersfeld (1992).

2. This anecdote comes from France's (1914), *Revolt of the Angels*.

3. Platonism includes notions of Platonic love and Platonic solids. Whereas the former refers to an idealized love, the latter refers to the five polyhedrons with perfect symmetry: the tetrahedron, the hexahedron (the cube), the octahedron, the (twelve-sided) dodecahedron, and the (twenty-sided) icosahedron. So perfect are these idealized objects that early astronomers, such as Kepler, assumed they determined the orbits of planets around the sun. Alchemists (chemistry's predecessors) assumed they described the shapes of the five elements: earth as the cube, wind as the octahedron, fire as the tetrahedron, water as the icosahedron, and the universe as the dodecahedron (Lundy, 2001).

4. Throughout this book, including its title, *mind* refers to all thought, whether conscious or subconscious. It does not imply the mind–body duality that Descartes espoused. As neuroscience makes clear, neural activity does not begin and end in the head. However, in contrast to some embodied perspectives, there is no claim that all thought is sensorimotor. Although sensorimotor activity provides the basis for most, if not all, of the mental actions described in this book, mental actions become coordinated with one another and abstracted from the particular contexts in which individual sensorimotor actions were first learned.

5. As acclaimed mathematician Francis Su explained, "the quest for truth predisposes the heart to the virtue of humility" (p. 487). Su is the author of *Mathematics for Human Flourishing*—a book that, like this one, highlights the humanity of mathematics, in themes of play, beauty, truth, justice, and love.

6. This quote comes from G. H. Hardy's, *A Mathematician's Apology* (1992, p. 151). Hardy considers his greatest accomplishment was to have worked with Srinivasa Ramanujan, whom Hardy "discovered" in 1913, in British-occupied India.

7. "For the mathematician it is, of course, tempting to believe in Ideas and to think of negative or imaginary numbers as lying in God's lap from all of eternity. But God himself has, since Gödel's theorem, ceased to be motionless. He is the living God, more so now than heretofore, because he is unceasingly constructing ever 'stronger' systems" (Piaget, 1970, p. 141).

8. For example, see Blum, Blum, and Shub (1982).

9. This view aligns well with "mathematical caring relations," as described by Amy Hackenberg (2010), and more generally to Nel Noddings' (2012) "caring relations in teaching."

10. You can find the NCTM complete position statement on access and equity on its website: nctm.org.

11. See, for example, Robert Berry's (2008) careful analysis of the mathematical identities of African American boys.

12. See Brown and Guy (2020) for a lively discussion of the triangle and its history.

13. As Sir Isaac Newton wrote in 1675, "if I have seen further, it is by standing on the shoulders of giants."

REFERENCES

Berry, R. Q. (2008). Access to upper-level mathematics: The stories of successful African American middle school boys. *Journal for Research in Mathematics Education, 39*(5), 464–488.

Blum, L., Blum, M., & Shub, M. (1982). *A simple secure pseudo-random number generator*. Electronics Research Laboratory, College of Engineering, University of California.

Brown, E., & Guy, R. K. (2020). *The unity of combinatorics*. Carus Mathematics Monograph Series (Vol. 36). Providence, RI: MAA Press.

France, A. (1914). *The revolt of the angels* (F. Chapman, Ed.). Retrieved from https://fooji.ir/wp-content/uploads/The-Revolt-of-the-Angels.pdf

Galileo, G. (1623). *The assayer* (S. Drake, Trans.). Retrieved January 21, 2021, from https://web.stanford.edu/~jsabol/certainty/readings/Galileo-Assayer.pdf

Glasersfeld, E. von (1992). A constructivist approach to experiential foundations of mathematical concepts. In S. Hills (Ed.), *History and philosophy of science in science education* (pp. 551–571). Kingston, Ontario: Queen's University.

Hackenberg, A. J. (2010). Mathematical caring relations in action. *Journal for Research in Mathematics Education, 41*(3), 236–273.

Hardy, G. H. (1992) *A mathematician's apology*. Cambridge: Cambridge University Press.

Lundy, M. (2001). *Sacred geometry*. New York: Walker Publishing.

Newton, I. (1675). "Letter from Sir Isaac Newton to Robert Hooke." *Historical Society of Pennsylvania*. Retrieved January 21, 2021, from https://digitallibrary.hsp.org/index.php/Detail/objects/9792

Noddings, N. (2012). The caring relation in teaching. *Oxford Review of Education, 38*(6), 771–781.

Piaget, J. (1970). *Structuralism* (C. Maschler, Trans.). New York: Basic Books (Original work published in 1968).

Su, F. E. (2017). Mathematics for human flourishing. *The American Mathematical Monthly, 124*(6), 483–493.

Valéry, P. (1935). *Cahiers* (Vol. 2). Paris: Gallimard.

1

What Is Number?

Suppose you are walking along a sidewalk and notice seven yellow bricks. How do you know they are bricks, that they are yellow, and that there are seven of them? *Yellow* and *brick* are subjective descriptors of perceptual experience. Determining whether something counts as a brick depends on language and cultural experience. Determining whether something is yellow depends on the way the cones and rods in our eyes transform electromagnetic waves into neural signals that we then interpret within some kind of color scheme, which again depends on language and cultural experience. Determining whether something is seven has a different nature.

The number 7 does not exist out there in the world for us to see. It begins in our minds, and then we project it out into the world.[1] Seeing 7 depends on the coordination of several mental actions. First, we have to notice something that we separate from the rest of our perceptual field. In the case of bricks in a sidewalk, the yellow ones stand out so we can easily focus attention on each of them. Second, we have to treat the objects on which we focus as if they were identical, even though no two things we see are truly identical. Each yellow brick is unique in some way (different hues of yellow, imperfections in their rectangular shapes, etc.), but we ignore idiosyncrasies for the moment and treat the bricks all the same, simply as objects to count. Finally, we put those identical things in one-to-one correspondence with our number sequence. For instance, we might point to each

DOI: 10.4324/9781003181729-2

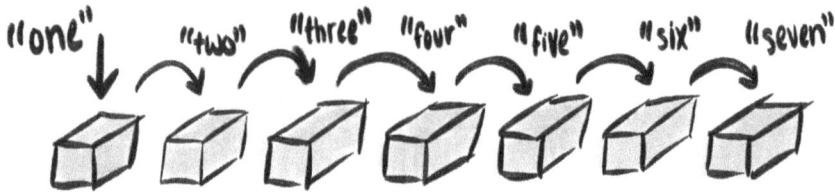

Figure 1.1 One-to-one correspondence between pointing acts and number words.

Source: © Eleanor Norton

brick as we recite, "One, two, three, four, five, six, seven" (see Figure 1.1). We might then view 7 as a property of the collection of bricks, when in fact, we have made it so through our own mental activity—even in making them a collection to count.

As adults, it's difficult to recall the laborious (and mostly subconscious) process of constructing number from our coordinated mental actions. We feel like we can simply see numbers, like we see colors, and so children should be able to see them too. We feel like we should be able to point to seven objects and say, "See? That's seven." It doesn't work that way because children have to learn to see numbers. They have to construct numbers before they can project them out into the world they perceive. "There is the mistaken belief that since we, as adults, can see mathematics in the blocks, the students will too. But the seeing requires the very construction the activity is intended to teach."[2]

For young children, counting requires great effort and concentration. Counting past 7 can be particularly tricky because the English word *seven* has two syllables, so children will sometimes point to the eighth item while reciting the second syllable in "se-ven".[3] Thus, children would lose the one-to-one correspondence between pointing acts and number words. Without this correspondence, reciting number words, "one, two, three . . . ," would be no more mathematical than reciting the alphabet, "A, B, C . . . ," or singing "*Do, re, mi* . . . " It would be nothing more than a memorized sequence of words—not counting.

ORDER AND CARDINALITY

The development of number comes in stages, beginning with one-to-one correspondence. Consider the case of my daughter, Eleanor, when she was 4 years old. Eleanor had been carefully counting out collections of chips when I asked her a question . . .

> **Me**: Is 9 bigger than 7?
> **Eleanor**: [slapping her hands to her face; see the left side of Figure 1.2] I'm thinking yes.
> **Me**: Why?

a   b   c

Figure 1.2 Eleanor comparing 7 and 9.

Source: © Eve Azano

Eleanor: Because one, two, three, four, five, six, seven [pointing to different places on an empty area of the table as she said each number word; see middle of Figure 1.2] and you haven't heard "nine" [raising hands in the air; see the right side of Figure 1.2]!

Eleanor's case teaches us a couple of things about numbers. First, learning how to count involves a careful coordination between pointing acts and number words in order to establish a one-to-one correspondence between them. We can see traces of that activity as children progress from needing physical objects to touch, to a focus on the acts of pointing themselves, to imagined acts of pointing, as indicated when a child looks up and nods for each number.[4] In Eleanor's case, she pointed to places on the table where she imagined the chips. It was the acts of pointing, and not the chips themselves, that were salient.

Second, learning to count is not the same as understanding numbers as quantities. Eleanor could easily count collections of 20 or more objects and had done so many times, but she did not know that 9 was bigger than 7 until she recited her number sequence again. For her, one number was bigger than another if it appeared later in the sequence. In other words, she compared numbers based on their order, not their cardinality.

In his studies of child development, Jean Piaget described the construction of number as an integration of order and cardinality, where cardinality refers to size or quantity.[5] As children integrate these two aspects of number, they begin to understand that one number is larger than another, not only because it follows in numerical order but also because that number contains the other. For example, 9 follows 7 and contains 7. This is not a simple matter for children to resolve.

Consider the case of my other daughter, Caroline. Starting when she was 5, I would ask her questions like the following: "How much is seven 1s?" The question may seem ridiculous to adults, but not to 5-year-old Caroline. She would always respond to such questions in the same way. In the case of seven 1s, she would count out seven fingers, raising them one at a time as she recited, "One, two, . . ., seven." Then she would point to each raised finger while reciting the same sequence of number words again! A year later, when I asked her about twenty 1s, she laughed

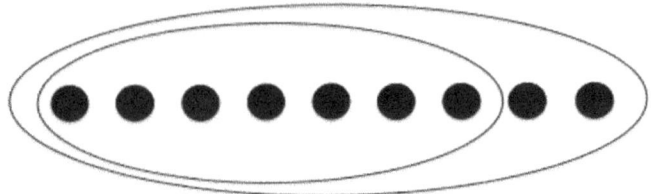

Figure 1.3 9 as a cardinality containing 7.

and said, "That's just 20!" Something had changed. Twenty 1s had become, for her, the same quantity as 20.[6]

As cardinalities, 7 is seven 1s, and 9 is nine 1s. Nine is bigger than 7 because the nine 1s in 9 include the seven 1s in 7. Between the ages of 5 and 6, Caroline seemed to develop an understanding of numbers as cardinalities. Figure 1.3 illustrates what this understanding entails.

Whereas order involves placing pointing acts in one-to-one correspondence with a number sequence, cardinality involves iteratively building up collections of 1s. As Caroline would later explain, seven 1s is 7 because 7 is 1 plus 1 plus 1 plus 1 plus 1 plus 1 plus 1. And as Eleanor later learned, 9 is bigger than 7 because it has two more 1s. With cardinality, children begin to understand numbers as quantities measured in units of 1. The number 1 becomes a unit of measure, like an inch or a meter.

UNITS OF UNITS

Once we understand numbers as measures, anything can be a unit for measuring, and whatever we choose as this initial unit becomes 1. This mental action of creating a unit is called *unitizing*. With the seven yellow bricks, we unitize instances of bricks, which we treat as identical. Their differences do not matter because they are simply objects to count, and it is the counting itself that matters. Likewise, we might unitize the entire collection of bricks or the distance between adjacent bricks. One (1) is simply a choice we make when determining how we want to measure collections or lengths.

Measuring amounts to iterating a unit of 1, where *iterating* refers to the mental action of making connected copies of an identical unit. For example, we might measure the distance on a track by iterating a meter or a yard. However, in mathematics, units do not need to refer to anything physical. A unit is simply the result of unitizing, producing a 1 that we might iterate to produce and measure other quantities.

As outlined in the previous section, constructing numbers as measures, or iterations of a unit of 1, takes years of experience. Manipulatives, such as fingers, chips, or blocks, can support those experiences by offering children a way of carrying out their mental actions on physical material. They can use the manipulatives to keep track of their activity, so they don't have to coordinate it in their minds all at once. This is a theme we will return to in Chapter 6.

Along the way, children will learn to count on from a given number to another. For example, they might determine the total number of objects from collections of eight objects and five objects by starting from 8 and counting on "9, 10, 11, 12, 13" while pointing to each of the five objects in the second collection (see Figure 1.4). They will even learn to double count, keeping track of the number of counts in the second collection: "9 is 1; 10 is 2; 11 is 3; 12 is 4; and 13 is 5." As you might imagine, this coordination of counts involves considerable effort for young children. However, these coordinations support children's construction of composite units, which are worth the time and effort.

Composite units are units made up of other units (units of units). For example, the numbers 7 and 9, as illustrated in Figure 1.4, are composite units: 7 is a unit made up of seven 1s, and 9 is a unit made up of nine 1s, including the seven 1s that make up 7. We might also build up 9 from composite units of 3: 9 as a unit made from three units of 3, each of which is three units of 1.

With composite units, we can measure the world not only in units of 1 but with all kinds of units. Moreover, we can relate those measurements. For example, consider the bars shown in Figure 1.5. Suppose the long bar is four times as long as the medium bar, which is three times as long as the short bar. You might

Figure 1.4 Counting on and double counting.

Source: © Eleanor Norton

Figure 1.5 Bars task.

understand the medium bar as a kind of composite unit, made up of three 1s. If so, you can determine the length of the long bar in units of 1 as well as units of 3.

Reflection

How many times will the short bar fit into the long bar?

A child might solve the bars task by first measuring the long bar with the medium bar. The child can then complete the task by repeatedly adding 3s: 3 + 3 + 3 + 3. This activity of repeated addition requires the child to segment her number sequence into 3s: 1, 2, 3; 4, 5, 6; 7, 8, 9; 10, 11, 12. Essentially, it is a continuation of counting, or counting by 3s. Although the task might look multiplicative to us, the child is treating it as an addition task and is using additive reasoning (repeated addition) to solve it. Multiplicative reasoning involves more.

PRESERVING AND TRANSFORMATION UNITS

Addition preserves units. In repeated addition, we chain together a sequence of composite units. These composite units are units that contain units of 1, so they involve two levels of units. Those two levels are maintained in addition. For example, in the case of four 3s, we can count in units of 3 or in units of 1. Either way, we maintain those two levels of units. In the end, we get a single sequence of numbers, segmented into units of 1 or units of 3 (see Figure 1.6).

Composite units also enable us to reason multiplicatively but only if we use them to transform units. Whereas additive reasoning involves a continuation of counting activity, ultimately in units of 1, multiplicative reasoning involves a transformation from units of 1 to composite units or vice versa. As noted in the previous section, 1 is whatever we choose to unitize, as our initial unit of measure.

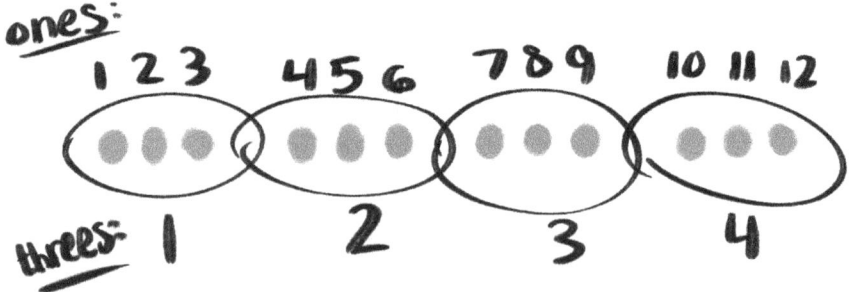

Figure 1.6 Four 3s.

Source: © Eleanor Norton

When we unitize a composite unit, we now have two units for measuring the world, as well as a way of relating those two levels of units. For example, the small bar in Figure 1.5 might be treated as a unit of 1, and three iterations of that unit create the medium bar as a composite unit of 3.

To solve the bars task multiplicatively, the child would need to transform the units of 3 into units of 1 all at once: If the long bar is four measures of the composite unit, 3, what does it measure in units of 1? There is a 3-to-1 exchange between the four units of 3 and the four units of three 1s, as represented on the left side of Figure 1.7. Each of the four composite units of 3 is transformed into three units of 1. It's as if the child has two number sequences—one for counting/measuring with the composite unit (3) and one for counting/measuring with 1s—and uses a 3-to-1 exchange to map between them (four units of 3 become four units of three 1s).

When seen in this light, multiplication is not repeated addition or a school-taught algorithm for computing products. It is the mental action of *distributing* the units of 1 within one composite unit across the units of 1 in the other composite unit. Said differently, we insert composite units into other composite units. In the preceding example, we inserted a composite unit of 3 into each unit of 1 within the composite unit of 4, to make four units of 3, or 12 units of 1. In general, the product $m \times n$ becomes n units of 1 inserted into each unit of 1 within m.

The mental action of distributing relates to the formal property of distribution, which relates the operations of addition and multiplication. Namely, the distributive property states that, if a, b, and c are whole numbers (or integers, fractions, real numbers, or even complex numbers), then $(a + b) \times c = (a \times c) + (b \times c)$. We can see this relationship represented geometrically in Figure 1.8.

The numerical relationship described by the distributive property relies on a coordination of units. We have to conceptualize a + b as a single unit composed of two other units, a and b, where a and b can be disembedded from the composite unit, a + b. Then we can understand the transformation of units represented by $(a + b) \times c$ as a transformation of units within each part: a and b. For instance, $(2 + 2) \times 3$ transforms the units of 1 within each unit of 2 into units of 3, so we have $(2 \times 3) + (2 \times 3)$. To make the connection to the mental action of distribution clearer, consider the following: $4 \times 3 = (1 + 1 + 1 + 1) \times 3 = 3 + 3 + 3 + 3$.

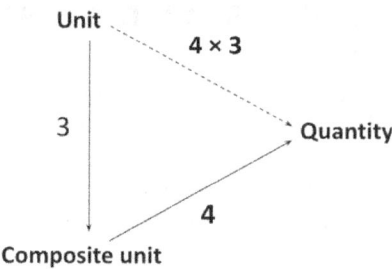

Figure 1.7 Triangle model of multiplication

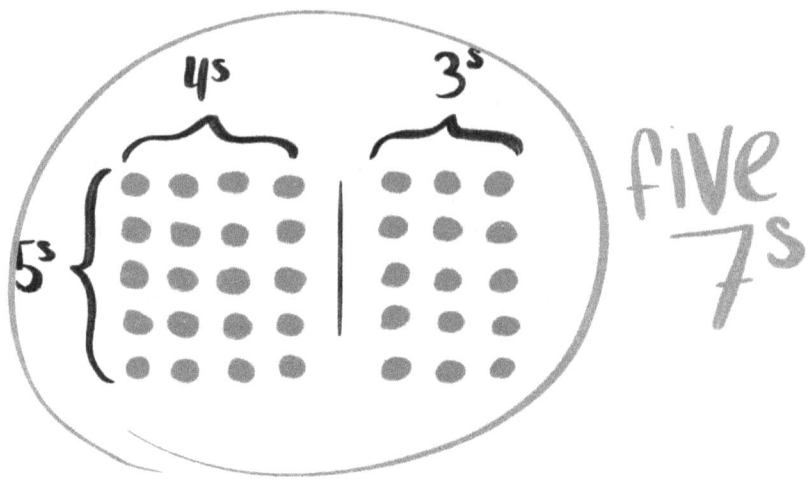

Figure 1.8 Five 7s.
Source: © Eleanor Norton

Therein, we also see a formal relationship between multiplication and repeated addition, even though students might conceptualize them differently: multiplication transforms units, and addition—even repeated addition—preserves units.

In Chapter 4, we will return to multiplication, in the context of fractions. There, we will also consider the commutativity of multiplication (m × n = n × m), which children do not take for granted. We might say that multiplication is conceptually noncommutative. This noncommutativity is marked by the terms *multiplicand* and *multiplier*, which refer to the distinct roles of the two terms in a product. In the product 4 times 3, for example, we treated 3 as the multiplicand; it describes the transformation of units. We treated 4 as the multiplier; it describes the number of units.

MATHEMATICS AND LANGUAGE

People learn to count in every language and in every culture. We count in different ways with different words, yet we can universally agree on the number of objects in a collection. In his studies of the Oksapmin people in Papua New Guinea, Geoff Saxe[7] found a particularly compelling example. Children in that culture are taught to count 27 places across their body, from their right thumb through their upper body to their left pinky (see Figure 1.9).

Addition and subtraction are accomplished by counting on or counting backward. For example, to subtract 6 from 13, an Oksapmin child would start by

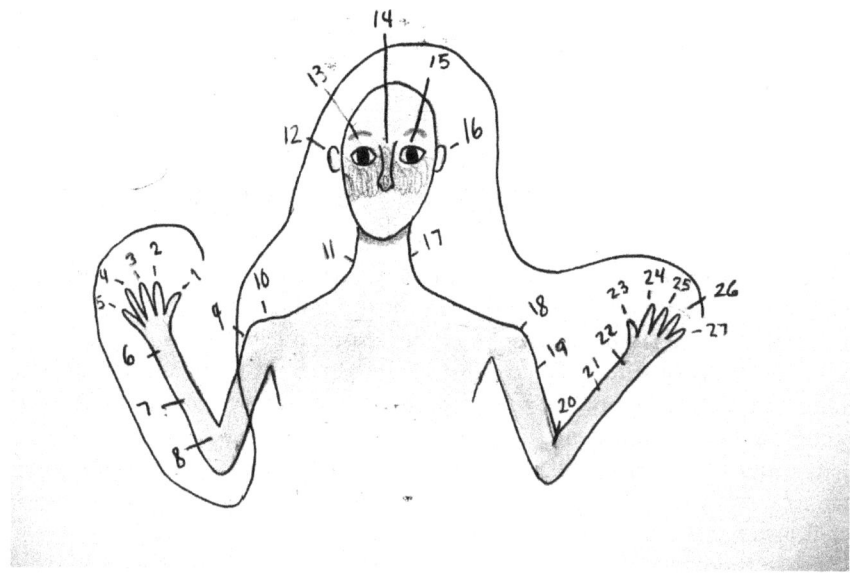

Figure 1.9 Counting in Papua New Guinea.

Source: Adapted from Saxe (1981). © Eve Azano

referring to her right eye (13) and give away six places while keeping track of the number of places given away: right eye is 1 (raising right thumb), right ear is 2 (raising right pointer), right side of neck is 3 (raising right middle finger), right shoulder is 4 (raising right ring finger), right upper arm is 5 (raising right pinky), and right elbow is 6 (pointing to right wrist), leaving the places up to the right forearm (7).

Reflection

Try solving the problem 22 − 9 using your body and the Oksapmin method.

The Oksapmin method may seem very different, but it relies on the same coordination of mental actions that children use when learning to count on their fingers. The Oksapmin people recite their number words in one-to-one correspondence with their pointing acts. The number words and the particular body locations may differ, but learning to count always involves a coordination between a sequence of number words and acts of pointing. Furthermore, our number systems develop as we begin to understand numbers through their cardinalities and not just their order. As we construct and iterate units of 1, we begin to understand that numbers appearing later in the ordered sequence contain the numbers that precede them. In the Oksapmin subtraction example, there is an implicit understanding that 13 contains 7 when the child gives away six places from the 13, leaving 7 places.

WHAT DO BABIES KNOW?

Psychological studies have suggested that humans, and some animals, have an innate ability to immediately apprehend small numbers, without counting. These studies show, for example, that infants can distinguish collections of two objects from collections of three objects. This ability is called subitizing. Because we, as adults, see those collections in terms of numbers, it is tempting to say that infants are conceptualizing numbers too.

In order to subitize, the perceived objects must be spatially separated, and infants can only subitize up to three or four objects. So, let's consider the possible spatial arrangements of one, two, and three objects, say, dots (see Figure 1.10). There are fundamental spatial distinctions between the possible arrangements of one dot versus two dots and two dots versus three dots—distinctions that have little to do with counting or conceptualizing numbers.[8]

There is only one possible spatial arrangement for one dot and only one possible spatial arrangement for two dots. The arrangement of two dots is distinguished from one dot by the separation between the dots (note the space between the two dots in the middle of Figure 1.10). This separation also introduces a line of symmetry between the two dots. There are two possible spatial arrangements for three dots: either they are collinear (line up with each other) or they are not. When the dots are collinear (as shown by the solid dots on the right side of Figure 1.10), there is a dot in the middle—something that can't happen with one or two dots. When the dots are not collinear, they suggest a bounded area (a triangular region, as shown by the hollow dots on the right side of Figure 1.10)—something else that can't happen with one or two dots. Thus, we see

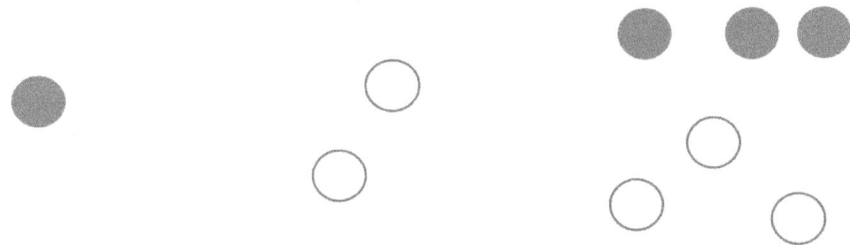

Figure 1.10 Spatial arrangements of one, two, and three objects.

that perceived spatial distinctions, such as separation and betweenness, can explain infants' ability to subitize.[9]

In addition to subitizing, psychologists have posited other innate mathematical abilities, such as magnitude comparisons and a mental number line. For example, babies and many animals are able to discern even large collections of objects if the sizes of the two collections are different enough, such as when comparing a collection of 20 dots to a collection of 80 dots. There are even studies suggesting that bees have a concept of 0 because they can be trained to find food under cards with no dots. However, as with subitizing, we can explain such observed behaviors without attributing an innate sense of number to babies or bees.[10] After all, numerical symbols and perceptual cues are not numbers; numbers are a product of mental actions that we learn to coordinate.

In general, we have to be careful about attributing our own knowledge to others. Just because we make distinctions between collections of dots based on counting and number does not mean that infants are making distinctions on that same basis. When we project our geometric structures into stars to see constellations, or when we project numerical patterns into a random array of numbers, the structures are ours, not the stars'. It is the same when teachers and parents work with children: although we can project our mathematical structures onto children as well as the heavens, children may see things very differently.

WHAT IS NUMBER?

What are numbers that we can know them, and who are we that we can know numbers?[11] Mathematicians have historically struggled with such questions. Euclid defined *number* as the "aggregate of several units"; Giuseppe Peano determined that *number* could not be defined; and David Russell introduced the following, circular definition: "a number is anything that is the number of some class."[12] Here, we have suggested that numbers are constructed, and therefore defined, through our own actions.

Returning to the yellow brick road, whether we count on our fingers or across our bodies as the Oksapmin people do, we count 7. We know there are 7, whether we call them "seven," "sept," "ilgob," or "beza." We know on the basis of our own coordinated activity. We can universally agree on results of counting and arithmetic because we refer to the same actions and the same coordinations thereof. We say $1 + 1 = 2$ because we understand addition as a continuation of counting; $1 + 1$ combines the two iterations of 1, which is precisely what we mean by (and how we arrive at) the number 2. After all, numbers do not exist in isolation but as part of a nested sequence in which numbers contain other numbers.[13]

The psychologist Ernst von Glasersfeld[14] provided a detailed analysis of addition for the case of $2 + 2 = 4$. He argued that we construct units of 1 from focused and unfocused moments of attention. We might focus on some element of our experience while ignoring others. As we encounter an element that interests us (e.g., a yellow brick), we focus on it, but then we are unfocused until we encounter that element again. In constructing units, the particular elements of interest do not matter, only the focusing and unfocusing of our attention. Each focused moment of attention (*) is bound by moments of unfocused attention (-), so that a unit would be represented by -*-. If that is our unit of 1, then, as two iterations of 1, 2 would look like this: -*—*-. 2 + 2 would combine the units of each 2, -*—*—*—*-, but when bound together, that is just -*-*-*-*- (i.e., 4).

SUMMARY

When we encounter the yellow bricks, we make them 7. First, we unitize the bricks (make them units to count). We focus on instances of yellow bricks and not on the space between the bricks. Next, we put these units in one-to-one correspondence with a sequence of number words. We can do this by coordinating acts of pointing with acts of reciting number words. Finally, we can take the seven bricks as a collection, a composite unit of 7 that contains seven units of 1.

As adults, we can coordinate these actions at once so that 7 appears to us as something that is out there, a property of the bricks and not our own mental activity. We take the actions for granted and hardly notice how we are coordinating them because we learned to do this at such an early age. It is only in observing the effortful activity of children that we are reminded that numbers are constructions. Reflecting on the following activities might raise awareness of—and appreciation for—that effort. As we will see, all mathematics develops in this way.

Activities

Activity 1: Look away from this text for a moment and find something to count (e.g., books, trees, pillows). How many are there?

Reflection: To answer this question, you must decide what counts as a unit (e.g., does a shrub count as a tree? Does a twin tree count as a single unit or two units?). You also need a method for keeping track of your counts so that you don't skip anything or count anything twice. Reflecting on this activity might remind us of some of the labor we experienced in learning to count, but you can take advantage of the results of your past labor: learning a number sequence and developing a structure for coordinating it with acts of pointing or noticing.

Activity 2: How many 7s are in twelve 21s? Can you solve this problem without multiplying 12×21?

Reflection: The challenge is not so different than the one Caroline faced when determining the value of seven 1s. Both tasks involve understanding numbers as quantities measured by another number, which serves as the unit of measure. In Caroline's task, 1 was the unit of measure. In the new task, 7 is the unit of measure.

Activity 3: In the following problems, letters are used to represent numbers: A is 1, B is 2, etc. Try to solve the problems without converting the letters back to numbers. For example, to solve M + C, you might start at M and count on from there by saying, "M + A is N; M + B is O; and so M + C is P."

    a. D + R
    b. N − J
    c. C × F
    d. X ÷ D

Reflection: To solve Task 3a, you might have taken advantage of commutativity, that D + R is the same as R + D, but this is not something children understand until they understand that R + D is a number composed of other numbers and that the same quantity, R + D, can be partitioned in different ways. To solve Task 3b, you take for granted that, because J comes before N, it must be contained in N, so you can count on from J. This way of reasoning relies on the nestedness of numbers, where smaller numbers are contained in larger numbers. To solve Task 3c, you might form a chain of Cs (or Fs) and then count by Cs (or Fs). This is the way children first learn to solve multiplication problems, through repeated addition, but is it really multiplication? Likewise, for Task 3d, is taking some number of Ds away from X really division or just repeated subtraction?

NOTES

    1. "Number is not a property of the objects which can be realized through the mere use of the senses, or impressed upon the mind by so-called external energies or attributes. Objects (and measured things) aid the mind in its work of constructing numerical ideas, but the objects are not number. Nor does the bare perception of them constitute number. A child, or an adult, may perceive a collection of balls or cubes, or dots on paper, or a bunch of bananas, or a pile of silver coins, without an idea of their number; there may be clear and adequate percepts of the things quite unaccompanied by definite numerical concepts. No such concepts, not clearly defined numerical ideas, can enter into consciousness till the mind orders the objects—that is, compares and relates them in a certain way" (McLellan & Dewey, 1896, p. 24).

    2. This quote comes from Wheatley (1992, p. 534), whose work complements the general thesis of this book: "mathematics is not out there in the real world but is the

learner's organizing activity" (p. 529). As the author explains, this idea extends to the use of instructional manipulatives (e.g., blocks) as well. Mathematics lives in manipulating, not in manipulatives.

3.   Clements and Sarama (2014) note this common coordination, between pointing acts and syllables, in their book *Learning and Teaching Early Math*. It demonstrates a developing one-to-one correspondence among early number learners—just not the one intended by their teachers.

4.   See work by Les Steffe (e.g., 1992) for a detailed elaboration on how children progress in their counting schemes. Catherine Ulrich (2015) provides an exquisite summary of this work.

5.   In the *Child's Conception of Number*, Piaget (1942) relied on clinical interviews with children to learn from them how humans construct numbers.

6.   "The child now finds this conservation so evident and necessary that he is sometimes visibly shocked to be asked such simple and 'childish' questions, when a year before he would have answered them quite to the contrary" (Piaget, 1980, p. 87).

7.   Saxe (1981) found that, although counting was "manifested in culturally specific ways" (p. 308), numerical development among Oksapmin children and children in the United States followed the same general progression.

8.   See Mandler and Shebo (1982) and von Glasersfeld (1982) for similar explanations of spatial distinctions between collections containing 1, 2, or 3 items.

9.   Although subitizing and pattern recognition are distinct from counting, students can rely on such patterns to support their development of counting. For example, Steffe (1992) shared the case of a 6-year-old student who learned to double count by relying on his imagined pattern for 5. It helped him keep track of the additional five counts as he counted on from 8, as described earlier: "9 is 1; 10 is 2; 11 is 3; 12 is 4; and 13 is 5." Mac-Donald and Shumway (2016) have described similar ways teachers can leverage subitizing to support their students' construction of number.

10.   Ulrich and Norton (2019) discuss the issue of over-attribution often found in research. It is human nature to project our ways of structuring the world onto others or onto the world itself. This is because our ways of structuring the world are the only ways we have of knowing it. Therefore, de-centering and understanding how others might structure their worlds require us to construct new knowledge. In this sense, we should be learning new mathematics from our children and students all the time.

11.   These questions have been paraphrased from the work of neuroscientist William McCulloch (1961), who claimed that the numbers 1 through 6 were perceptible numbers, which "we share with the beasts" but that numbers 7 and higher had to be counted as only humans can.

12.   This definition appears in Russell (1993, p. 19). To be fair, Russell did elaborate by referring to a one-to-one relation between classes. However, as Moreno (1974) explained, Russell's attempts to reduce mathematics to the logic of sets and classes ultimately fail because formal logic contains only quantifications like "for all" and "there exists." Moreover, as Inhelder and Piaget have demonstrated, children construct both number (Piaget, 1942) and logic (Inhelder & Piaget, 1964) through their construction of reversible and composable (logico-mathematical) operations.

13.   See Steffe (1992) or Ulrich (2015) for a description of nested number sequences.

14.   von Glasersfeld's (1981) "attentional model" has a striking similarity with Saint Augustine's explanation for the elusiveness of the number zero: "nevertheless, we have no sensation but the privation of sensation. For example, when the vision of the eyes passes from sensation to sensation, it sees darkness only when it begins not to see" (as quoted in Moreno, 1974).

REFERENCES

Clements, D. H., & Sarama, J. (2014). *Learning and teaching early math: The learning trajectories approach*. New York: Routledge.

Glasersfeld, E. von (1981). An attentional model for the conceptual construction of units and number. *Journal for Research in Mathematics Education, 12*(2), 83–94. doi:10.2307/748704

Inhelder, B., & Piaget, J. (1964). *The early growth of logic in the child*. New York: Norton.

MacDonald, B. L., & Shumway, J. F. (2016). Subitizing games: Assessing preschoolers' number understanding. *Teaching Children Mathematics, 22*(6), 340–348.

Mandler, G., & Shebo, B. J. (1982). Subitizing: An analysis of its component processes. *Journal of Experimental Psychology: General, 111*(1), 1.

McCulloch, W. S. (1961). What is a number, that a man may know it, and a man, that he may know a number. *General Semantics Bulletin, 26*(27), 7–18.

McLellan, J. A., & Dewey, J. (1896). *The psychology of number and its applications to methods of teaching arithmetic* (Vol. 33). New York: Appleton Press.

Moreno, A. (1974). Bertrand Russell's concept of number. *Angelicum, 51*(1), 88–110.

Piaget, J. (1942). *The child's conception of number*. London: Routledge & Kegan Paul.

Piaget, J. (1980). *Adaptation and intelligence: Organic selection and phenocopy*. Chicago: University of Chicago Press.

Russell, B. (1993). *Introduction to mathematical philosophy*. New York: Dover.

Saxe, G. B. (1981). Body parts as numerals: A developmental analysis of numeration among the Oksapmin in Papua New Guinea. *Child Development*, 306–316.

Steffe, L. P. (1992). Schemes of action and operation involving composite units. *Learning and Individual Differences, 4*(3), 259–309.

Ulrich, C. (2015). Stages in constructing and coordinating units additively and multiplicatively (Part 1). *For the Learning of Mathematics, 35*(3), 2–7.

Ulrich, C., & Norton, A. (2019). Discerning a progression in conceptions of magnitude during children's construction of number. In *Constructing Number* (pp. 47–67). Cham, Switzerland: Springer Nature.

von Glasersfeld, E. (1982). Subitizing: The role of figural patterns in the development of numerical concepts. *Archives de Psychologie, 50*(194), 191–218.

Wheatley, G. H. (1992). The role of reflection in mathematics learning. *Educational Studies in Mathematics, 23*(5), 529–541.

2

Reflections Upon Reflections

Symmetry refers to a transformation of a figure that leaves the figure looking exactly as it did before, occupying the same position in space. With reflective symmetry, we can think about the transformation as flipping a figure over a line. Reflective symmetry is a psychological primitive, meaning that even infants recognize it. Psychological studies indicate that, in their first year, children recognize and often prefer images with vertical symmetry, like faces.[1] By relying on symmetry, you probably have no difficulty imagining the entire face of the tiger whose half-face is shown in Figure 2.1.

Reflection

Try to become aware of the mental action you perform when imagining the other half of the tiger.

Symmetry is a mental action that we can perform with little or no effort, but it is powerful nonetheless. Its power owes to two key properties of the mental action, *reflecting*:

- Reflections are reversible;
- Reflections can be composed with one another.[2]

DOI: 10.4324/9781003181729-3

Figure 2.1 Reflective symmetry of the tiger.

Source: © Eleanor Norton

Here, we consider those two properties and their mathematical power.

REVERSIBILITY

What happens when you perform a reflection twice? For example, what happens when you reflect the half-tiger (see Figure 2.1) over its line of symmetry and then reflect it over that line again? The net effect has no effect on the original image; we're back to the same half-tiger. This is what we mean by reversibility: a mental action is reversible if we can perform another mental action that undoes it. In the case of reflection, the action and its inverse (the reverse action) are the same. In other examples that we consider, one mental action is undone by a different mental action so that they form a pair of inverse actions.

COMPOSABILITY

What happens when you compose two reflections? In other words, if you perform a reflection and then perform another reflection, what is the combined action? If the two reflections are the same, the answer is easy: the combined action is no action at all, as we just saw. But what happens when the two reflections are not the same—when there are two different lines of reflection? Consider the examples shown in Figure 2.2.

There are two cases to consider: when the lines are parallel (see the left side of Figure 2.2) and when they intersect (see the right side of Figure 2.2). The first case is similar to reflecting over the same line twice, except the two lines are shifted apart by some distance, d. To see exactly what happens, it is helpful to investigate what happens when we perform the pair of reflections on a particular figure, such as the turkey face.

Reflection

Imagine what would happen to the turkey (as shown in Figure 2.2), if the two lines of reflection were moved.

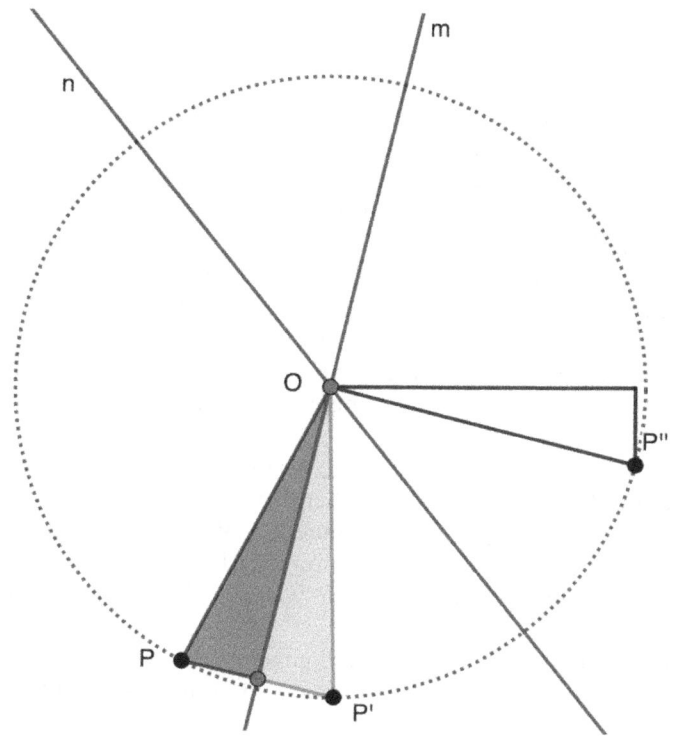

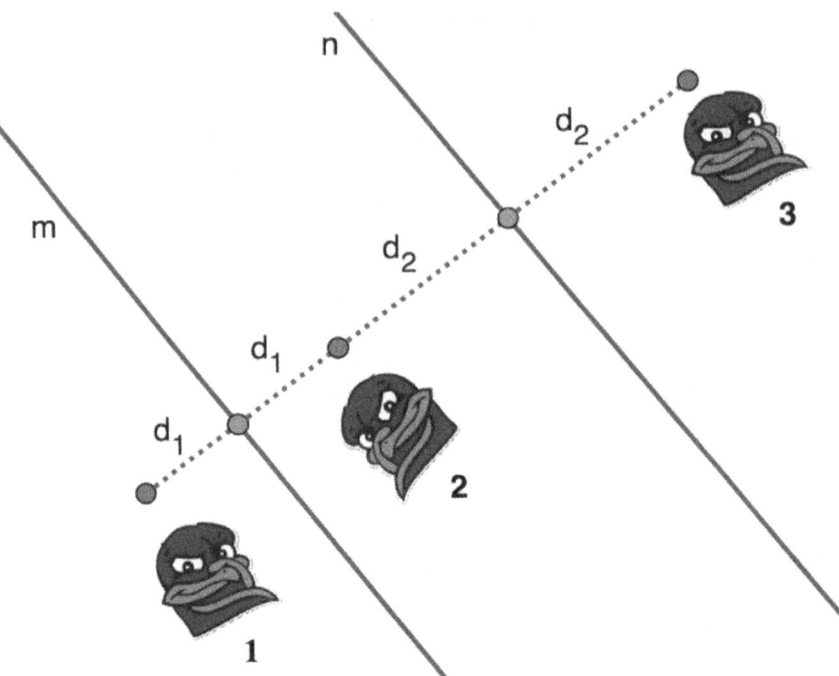

Figure 2.2 Composing reflections about two different lines, intersecting (top) and
parallel (bottom).

Notice that the first reflection (over line m) gives the mirror image of the turkey's face, switching its left eye and right eye. However, just like reflecting over the
same line twice, the second reflection (over line n) returns the face to its original
orientation. The only difference is that the face has been shifted by a distance of
2d. Breaking up d into d_1 (the distance from the original face to line m) and d_2
(the distance from the original face to line n), we can see why. Just consider how
each reflection reproduces those distances on the other side of the line, and then
consider their net effect in shifting the original face to the final face. Although
the face is a useful figure in figuring this out, keep in mind that what is essential
is your coordination of the reflections themselves—the coordination of your own
mental actions.

The second case is more challenging, meaning that the two mental actions
become more difficult to coordinate. Whereas we can easily perform the reflection over line m (see the right side of Figure 2.2), it becomes difficult to imagine
the net effect after performing the reflection over line n. We might ask ourselves,
What changes and what remains the same when performing the two reflections
in sequence?

Again, it is helpful to consider the net effect of the two reflections on a particular figure to keep track of transformations and invariances (that which remains

the same under the combined transformation). Let's start with the dark triangle shown on the right side of Figure 2.2. Note that it is a right triangle with one side on line m. Note also that it has one vertex at O, which is the point where lines m and n intersect. Finally, note that O remains invariant under both reflections (because it is on each line of reflection) and is therefore invariant under the combined transformation. All other points will be affected by one reflection or the other, or both.

Take, for instance, point P, which is another vertex on the dark triangle. Using our primitive mental action of reflection, over line m, we see that P is transformed to P′ and that the dark triangle is reflected over line m onto the lighter gray triangle. We can also see that all the angle measures and side lengths of the right triangle remain invariant. The same is true when we reflect the lighter gray triangle across the second line, n; we get the white right triangle with all the same angle measures and side lengths as the original triangle. Moreover, the triangle is oriented the same as the original, with the right angle on the right-hand side.

What does this tell us about the net effect of composing the two reflections? The point, O, doesn't move, and although the triangle doesn't change shape, the rest of it does move. Specifically, consider what happens to the point, P. In being transformed to P′ and then to P″, its distance to O does not change, because the triangle side, OP, didn't change length. However, it rotated around O, and we can see by how much if we look at the angles around O. Namely, it's twice the angle between the two lines, m and n. Furthermore, we can see that there is nothing special about the point P; it could have been any point in the plane. Thus, all points in the plane are rotating counterclockwise, around O, by an angle that is twice the angle between the two lines of reflection.

We are beginning to see the mathematical power of reversible and composable actions. From the primitive action of reflection, we can generate every possible rotation and translation of the plane. Moreover, this mental action—this mathematical power—is yours. It did not originate in a geometry textbook, in the fabric of the universe, or in the mind of some ancient Greek mathematician. We can reflect tigers and rotate the universe at our will. Mathematical power resides in every individual, on the basis of their own mental actions, should they choose to exercise it. Let's develop that power further.

COMPOSING TRANSFORMATIONS

We have seen that two reflections form a rotation (when the lines of reflection intersect) or a translation (when the lines of reflection are parallel). So, what happens when we compose two rotations or two translations? What happens when we compose a rotation and a reflection, or a translation and a rotation? In answering these questions, we will build up a group of transformations until we no longer get any new kinds of transformations.

The word *group* has a familiar meaning, but it also has a particular mathematical meaning.[3] Roughly, a *group* is a collection of things (in the present case, transformations) whose structure describes how those things can be reversed and how they can be combined with one another to produce other things in the collection. In other words, groups are sets of reversible elements that are closed under composition. This rough definition fits well with our characterization of mathematics as the coordination of reversible and composable actions.

Swiss psychologists Jean Piaget and Barbel Inhelder described a set of spatial transformations whose coordination is critical to children's development—the group of displacements in space.[4] Babies aren't born with a conception of space; they have to construct it through their own actions. At first, the world appears like a movie screen of blurred colors that come and go, as if from nowhere and into nothingness. This is why peekaboo is such an entertaining game!

As they begin to crawl, infants learn that they can control what they see through their movements. Moving one way, the movie changes, and then moving back, it returns to the same scene. These movements correspond to translations— the same kinds of translations we've been considering, except now they occur in three-dimensional space. More accurately, we might say that the group of these translations creates the three-dimensional space that we, as adults, take for granted.

Returning to the plane, we see that every translation has an inverse translation and that composing two translations yields another translation. Thus, translations form a group in themselves. We also know that two reflections form either a translation or a rotation. Now, let's consider what happens when we combine two rotations. When the two rotations occur around the same point, A, the answer is simple: we get another rotation about that point, where the two angles of rotation are added together. But what happens when we compose rotations about two different points, A and B (see Figure 2.3)?

Here we have a clockwise rotation about point A of 45 degrees followed by a clockwise rotation about point B of 75 degrees. To determine what happens when we compose the two rotations, we can focus on how the turkey is transformed. We can make the problem simpler by decomposing each rotation into a pair of reflections—the most basic mental actions.

The rotation about point A can be broken into a reflection over the dashed line through point A (the angle bisector of the 45-degree angle at point A) followed by a reflection over the line through points A and B. Likewise, the rotation about point B can be broken into a reflection over the line through points A and B followed by a reflection over the dashed line through point B (the angle bisector of the 75-degree angle at point B). In both cases, the pair of reflections generate the corresponding rotation because they intersect at the point of rotation (points A and B, respectively), and the angle between the lines of reflection is half of the respective angle. Also, note that we carefully chose the pairs of reflections so that they would have a line of reflection in common—the line through points A and B.

What happens when we put the two pairs of reflections all together? We have reflections over the four lines where the middle two lines are the same. But if we put the middle two reflections together—the reflections over the same line (the

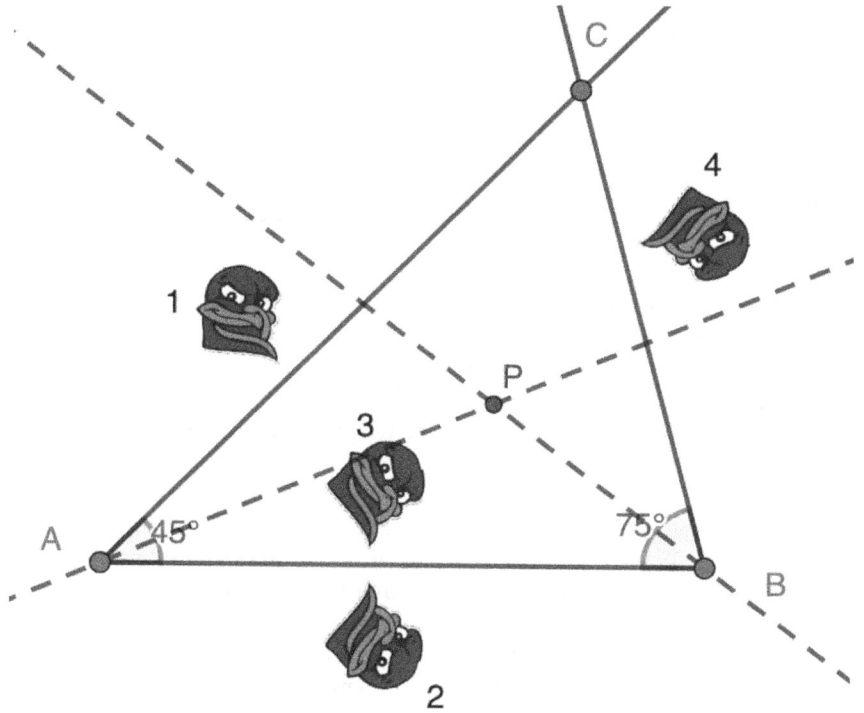

Figure 2.3 Composing two rotations about different points.

line through points A and B)—they have no net effect because the reflection over any line is its own inverse. So, we are left with reflections over the pair of dashed lines. We already know that the composition of a pair of reflections forms a rotation about their point of intersection (point P) that is twice the angle between the lines of reflection. Thus, the composition of a pair of rotations, even if those rotations are about different points, forms another rotation.

Reflection

Reflect on the four reflections in Figure 2.3. Follow the resulting effect on the turkey after each reflection, starting with the turkey with two blue dots beneath it and the reflection over the dashed line through point A. Can you visualize each reflection, and can you visualize the various pairs of reflections as rotations about point A, point B, and, finally, point P?

We began this chapter by considering reflections, and we have seen that composing two reflections yields a translation or rotation. In fact, we can break down

Table 2.1 Transformations of the plane[5]

	Reflection	*Translation*	*Rotation*
Reflection	Translation or rotation		
Translation		translation	
Rotation			rotation

any translation or rotation into a pair of reflections. In that sense, we can say that the mental action of reflection generates all possible translations and rotations in the plane. Composing two translations yields a translation, and we just saw that composing two rotations yields a rotation. Table 2.1 summarizes what we know so far and introduces a question: what happens when we compose reflections, translations, and rotations with one another?

Composing a reflection with a translation yields a transformation called a glide reflection; as the name suggests, this is just a reflection that then glides over (i.e., is translated, or shifted). We can also rotate reflections and translate rotations. In general, we begin generating all kinds of transformations that preserve the angles and lengths within geometric figures. Although a turkey face might get moved around in space, it is still the same figure with respect to its internal relationships (e.g., the eyes stay the same distance apart). Such transformations are called isometries. All these isometries began from composing reflections.

THE GROUP OF ISOMETRIES

Ultimately, reflections generate all isometries of the plane. To be specific, an isometry is a transformation of the plane that maintains the distance between any pair of points. For example, if we take a triangle in the plane and perform an isometry of the plane, the triangle might change positions, but it will have the same angle measures and side lengths as before. To see all the ways that might happen, we can focus on what might happen to its three vertices. Specifically, consider the three vertices of triangle ABC, as shown in Figure 2.4.

First, we can move vertex A anywhere in the plane, via a translation, to vertex A′. However, vertex B has to remain the same distance from A as it was before the isometry: the distance from B′ to A′ must be the same as the distance from A to B. So, once the position of A′ is determined, B′ must be somewhere on the circle centered at A′ (see the right side of Figure 2.4). In other words, after triangle ABC is translated, B can be transformed further but only in such a way that it is rotated around point A′. With the position of B′ determined, there are only two possibilities for C′, because C′ has to remain a fixed distance from A′ and B′. Thus, after C is translated (along with A and B) and rotated (along with B), it can be reflected to

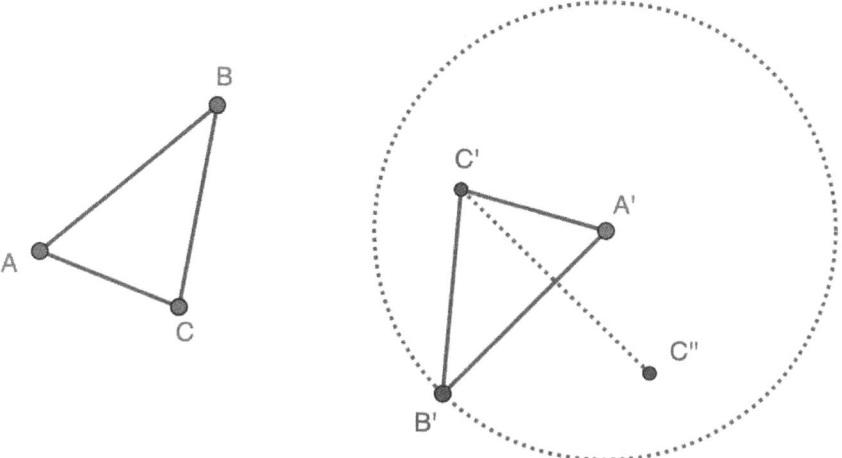

Figure 2.4 Isometries of the plane.

C″ or remain at C′. To summarize, every isometry of the plane can be described by a translation, followed by a rotation, possibly followed by a reflection.

Relying on the mental action of reflection and our coordination of various reflections to generate translations and rotations, we can show even more. We can show that every isometry of the plane is the composition of at most three reflections. After all, a translation is the composition of two reflections, and a rotation is the composition of two reflections. So, we already know a translation followed by a rotation followed by a reflection is the composition of five reflections. We can get this number down to three if we choose the pairs of reflections that represent the translation and rotation to have one reflection in common (like we did in composing rotations, in Figure 2.3). Consider the reflections shown in Figure 2.5.

Reflection

What is the result of reflecting the plane over line l and then reflecting it over line m? What is the result of reflecting it over line m and then line n? Finally, what is the result of reflecting it over line l and then line n?

By reflecting on your reflections, you should notice that the first pair of reflections yields a translation and that the second pair yields a rotation about point P. The third pair of reflections is the composition of these two pairs of reflections. Symbolizing each reflection with an r and a subscript indicating the line of reflection, we have the following composition: $r_l r_m r_m r_n = r_l r_n$. It results in a rotation about point Q. Thus we find that the composition of a rotation and a translation yields a rotation about some other point. This means every isometry of the plane

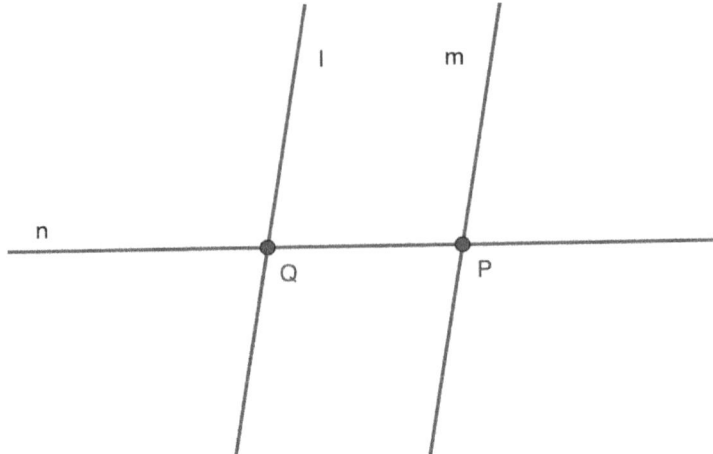

Figure 2.5 Reflecting on reflections.

is the composition of at most three reflections. If the isometry is not a reflection, translation or a rotation, it is a translation or rotation followed by a reflection.

Now that we have described every isometry as a composition of reflections, we can describe all possible compositions of isometries. In particular, we know that the composition of any two isometries is ultimately a composition of reflections. We know that every composition of reflections is an isometry, so we also know that the composition of any two isometries is another isometry. Furthermore, we can reverse any isometry by decomposing it into (at most three) reflections and reversing their order.

Reflection

The isometry represented by the composition $r_l r_m r_n$ has the composition of $r_n r_m r_l$ as its inverse. How might you show that?

Summarizing, isometries are closed under composition (combining any two of them yields another isometry), and every isometry has an inverse (we can reverse them by reversing the order of reflections that comprise them). These are two of the four properties that define a mathematical group. A third property is satisfied if we accept doing nothing to the plane as an isometry of the plane (think of it as composing a reflection with itself, if you'd like). We might call this the identity isometry because it leaves everything identical to the way it was before.

Isometries form a group—one of the simplest structures in mathematics. These structures underlie such diverse domains as adding integers, multiplying fractions, permuting (rearranging) letters, and composing functions. In essence, groups describe how we can compose and reverse elements in a set, as we have done with the set of reflections. Although group theory was not formally developed until the

19th century, we will see that group-like structures are deeply rooted in human psychology and undergird all mathematical development. We will return to this idea in Chapter 3, where we formally define groups.

TRANSFORMATIONAL GEOMETRY

When we focus on the mental actions that undergird geometry (e.g., rotations and reflections), the subject becomes much more intuitive. Consider the geometry theorem stating that the sum of angles in a triangle is 180 degrees. For example, the triangle shown in Figure 2.6 has angles of 30, 60, and 90 degrees, which add up to 180 degrees.

By definition, all triangles have three sides connected at vertices, where the angles are formed. In fact, we can define an angle as a rotation between two sides, as one side rotates (around the common vertex) to the other side. Note that this dynamic definition of angle stands in contrast to the static definitions usually found in textbooks. Many high school textbooks define an angle as two rays (one-directional lines) starting from the same vertex. Our dynamic definition of angle emphasizes the mental action of rotation, and it makes a proof of the theorem more intuitive.

Figure 2.7 shows four copies of the same triangle, with angles A, B, and C. The measure of angle A is the amount of rotation needed to move the bottom side to the left side (see Figure 2.7a). Note that the bottom side has an arrow to help us keep track of the rotations. In particular, when rotated by angle A, the bottom side would point in the direction shown in Figure 2.7b (ignoring the change in length). When rotated by angle B, the tail of the arrow would swing around so that the arrow would point in the direction shown in Figure 2.7c. Finally, when rotated by angle C, the arrow would face the direction shown in Figure 2.7d. How far has the arrow turned?

Focusing on the mental action of rotation (rotating each side of the triangle to the next side of the triangle) enables us to understand why the sum of angles

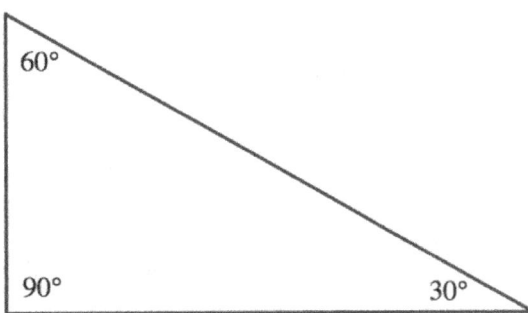

Figure 2.6 Sum of angles in a triangle.

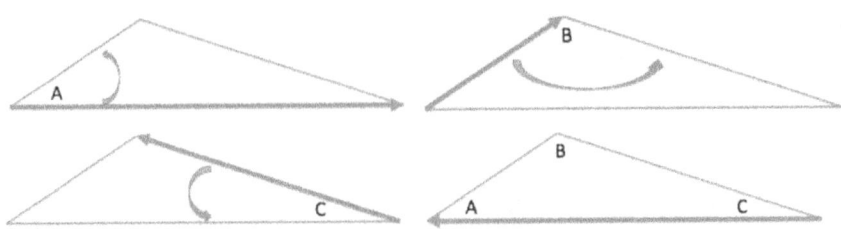

Figure 2.7 Proof about the sum of angles in a triangle.

in any triangle must be 180 degrees. In coordinating the three rotations, we find that the arrow shown in Figure 2.7a turns to the opposite direction, as shown in Figure 2.7d—a 180-degree rotation. This proof does not rely on measuring angles in particular triangles, and it does not rely on formal axioms (discussed in Chapter 5). We can agree on the universal truth of the theorem, not because it is written into the fabric of the universe or because mathematics textbooks insist on it but because it refers to a sequence of mental actions that we can all perform.

Figure 2.8 illustrates the coordination of these three mental actions (rotations), which may be easier to imagine as rotations of the three lines containing the three sides of the triangle. The angle shown at vertex *A*, on the left side of the figure, is represented as angle CAB on the right side of the figure. Note that it rotates the line at the bottom of the triangle, about vertex A. Next, the angle at vertex B (angle ABC) rotates the line through the left side of the triangle. Finally, the angle at vertex C (angle BCA) rotates the line through the right side of the triangle, thus completing the 180-degree rotation.

Transformational geometry refers to arguments like these: arguments based on mental actions we might perform (composing three rotations), rather than properties of static figures (two rays meeting at a point form an angle). In Chapter 3, we use transformational geometry to classify quadrilaterals, based on their symmetries. We describe these symmetries as groups of transformations, extending a program begun by mathematician Felix Klein in the 1800s.

SUMMARY

We have seen the power of a single mental action, reflection, in generating an entire group of transformations of the plane. Rotations, translations, glide reflections, and every isometry of the plane arise from the coordination of reflections. As a primitive mental action, reflection offers a mathematical power that is available to all of us when we exercise its coordination. By reflecting on reflections, we become aware of space as a group of transformations that we generate with our own minds.

By focusing on space as a product of our own psychology, we make geometry more intuitive. That's because mental actions, like rotations, are the basis for our

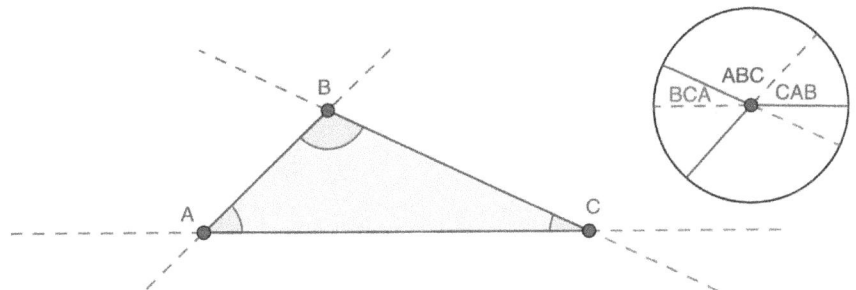

Figure 2.8 Triangle angle sum.

construction of geometric objects, such as triangles, as well as geometric space itself. We saw, for example, that the angles in a triangle must sum to 180 degrees, not because of a list of arguments and theorems in a textbook but because of the way we can compose angles, as rotations, in our own minds. In fact, as we will see in Chapter 5, formal arguments are attempts to clearly communicate the mental actions that have always guided mathematicians' proofs of theorems.

Students sometimes find geometry appealing because they can visualize its objects of study: shapes. They can use figures (diagrams, drawings, or geometric constructions) to reason through relationships within shapes. For example, the diagrams shown in Figures 2.7 and 2.8 supported our arguments about angles within triangles. It would be difficult to reason through the arguments without such figures. Thus, we might say that the arguments involve "diagrammatic reasoning."[6] However, figures are secondary to the transformations they represent.[7]

In the case of Figure 2.7, the diagrams indicate the transformation of one side of the triangle to an adjacent side, by an angle rotation. The angle rotations themselves, rather than the diagrams that represent them, are salient for the argument. However, without the diagrams, it would be difficult to keep track of the three rotations and their cumulative effect on the initial side (a 180 rotation). In other words, we use figures to help us coordinate the various mental actions that form and transform geometric objects, but mathematics lives in the dynamic actions/transformations that we perform rather than the static figures that represent them.

Activities

The following activities include figures, and the figures might help you visualize and keep track of your mental actions. As you work through the activities, keep in mind that you are coordinating of your own mental actions rather than the figures themselves.

Activity 1: What transformation do you get when you compose the four reflections shown in Figure 2.9, first reflecting over line k, then line l, then line m, and then line n?

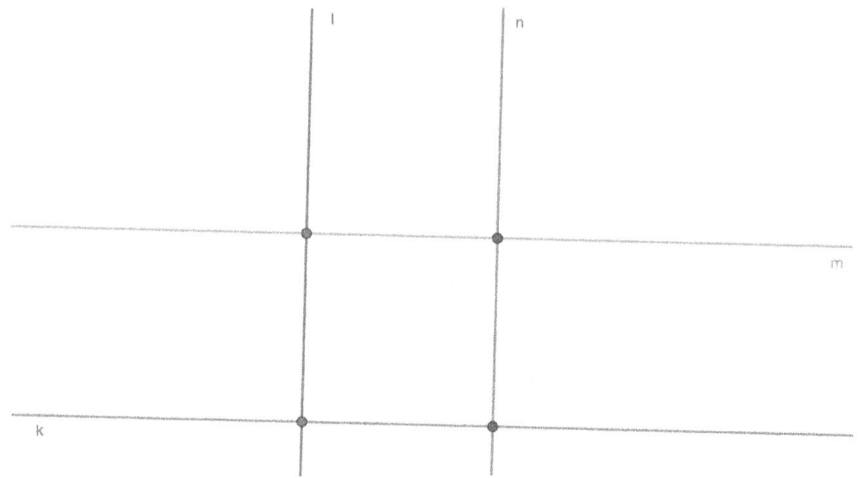

Figure 2.9 Composing four reflections

Hint: to keep track of the reflections, it might help to draw a simple figure (an asymmetric figure like the side of a face) and see what happens to it after each successive reflection.

Activity 2: Draw two lines of reflection whose combination would result in the translation on the left side of Figure 2.8, and two lines of reflection whose composition would result in the rotation (about point O) on the right side of Figure 2.10.

Figure 2.10 Decomposing translations and rotations.

Activity 3: Use the diagram in Figure 2.11 to argue that the interior angles in a pentagon sum to 540 degrees. Can you generalize this argument to all polygons?

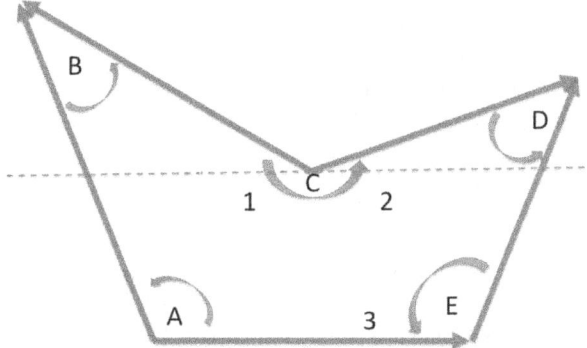

Figure 2.11 Angle sums in pentagons.

NOTES

1. For example, Bornstein, Ferdinandsen, and Gross (1981) showed that four-month-old children recognized symmetry and that, by twelve months of age, children preferred vertically symmetrical figures to horizontally symmetrical figures.

2. Piaget (1970) characterized logico-mathematical operations as mental actions that are reversible and composable.

3. We will introduce the formal definition of a group in Chapter 3.

4. "The 'intuition' of space is not a 'reading' or apprehension of the properties of objects, but from the very beginning, an action performed on them" (Piaget & Inhelder, 1967, p. 449).

5. On one hand, Dehaene, Izard, Pica, and Spelke (2006) found that children develop geometric concepts like perpendicularity intuitively, outside of schooling. On the other hand, Portnoy, Grundmeier, and Graham (2006) found that formalizing related transformations was challenging for college students.

6. Hoffmann (2004) introduced diagrammatic reasoning—manipulating and performing constructions within diagrams—as a productive means for forming conjectures in geometry.

7. Tall, Thomas, Davis, Gray, and Simpson (1999) make a distinction between the square as a figural object and as an operative object. Unlike the number, 5, which is only operative, the square can be taken as a perceptual unit. The symbol, 5, does not represent operative aspects of the number, but a square figure can represent operative aspects of the square as a mathematical object (e.g., four right angles and four equal sides). This relates to the Greek idea of Platonic objects, which can be represented (albeit imperfectly) in drawings.

REFERENCES

Bornstein, M. H., Ferdinandsen, K., & Gross, C. G. (1981). Perception of symmetry in infancy. *Developmental psychology, 17*(1), 82.

Dehaene, S., Izard, V., Pica, P., & Spelke, E. (2006). Core knowledge of geometry in an Amazonian indigene group. *Science, 311*.

Hoffmann, M. H. (2004). How to get it: Diagrammatic reasoning as a tool of knowledge development and its pragmatic dimension. *Foundations of Science, 9*(3), 285–305.

Piaget, J. (1970). *Structuralism* (C. Maschler, Trans.). New York: Basic Books (Original work published in 1968).

Piaget, J., & Inhelder, B. (1967). *The child's conception of space* (F. J. Langdon & J. L. Lunzer, Trans.). New York: Norton (Original work published in 1948).

Portnoy, N., Grundmeier, T. A., & Graham, K. J. (2006). Students' understanding of mathematical objects in the context of transformational geometry: Implications for constructing and understanding proofs. *Journal of mathematical behavior, 25*(3).

Tall, D., Thomas, M., Davis, G., Gray, E., & Simpson, A. (1999). What is the object of the encapsulation of a process? *The Journal of Mathematical Behavior, 18*(2), 223–241.

3

Getting With the Program

Chapters 1 and 2 illustrate how we construct numbers and space through the coordination of our own mental actions. This approach to mathematics, as a product of our own psychology, originated with Swiss psychologist Jean Piaget. However, it has roots in philosophy[1] and mathematics itself. Specifically, in the late 19th century, famed German mathematician Felix Klein established a program for classifying various branches of geometry based on how they transform space.[2] Here, we will examine that program, explicitly connect it to psychology, and extend it to all of mathematics.

The Erlangen program depends upon a foundational structure in mathematics, called a group. Chapter 2 provides some indication for how space is constructed as a group—a group of displacements that children begin to tacitly construct in infancy.[3] In Chapter 4, we will see how operations on numbers might also form groups. Groups describe ways of structuring formal operations, such as addition and multiplication, as well as the less formal mental actions that guide our intuitions about space and number.

GROUP THEORY

Chapter 2 introduced the idea of a mathematical group. We define it here as a set of composable elements with the following four properties:

1. If you compose (combine) any two elements in the set, you get another element in the set. Formally, we say that the set is closed under composition (denoted here by "∘"): if a and b are in the set, then so is a∘b.
2. There is an element in the set, called the identity, that has no effect on other elements when combined with them. If we call the identity 'i,' then for any element, a, in the set, a∘i = i∘a = a.
3. Every element in the set has an inverse element in the set that reverses it so that their combination has no net effect. If a is in the set, then there is an inverse element, say a^{-1}, in the set such that $a∘a^{-1} = a^{-1}∘a = i$.
4. If you combine three elements in the set, you can combine the first two elements first and then combine that result with the third element, or the last two elements first, and then apply that result to the first element; the net effect is the same either way. Formally, we call this property associativity: if a, b, and c are elements of the set, then (a∘b)∘c = a∘(b∘c).

The preceding definition fits our characterization of mental actions very well if we consider doing nothing as a trivial mental action (the identity; property 3). In particular, if our mental actions are composable, we can combine them to form other mental actions, which will be in the group of mental actions (property 1). Also, because mental actions are reversible, they satisfy property 2. All that remains is to ensure property 4: associativity. This property is satisfied whenever mental actions transform a space (whether it be a geometric space or a space of numbers), thus mapping one version of the space to another version of that space. The composition of mental actions would then be a composition of mappings, which are always associative because each mapping picks up where the prior mapping left off.[4]

Take Piaget's group of displacements (introduced in Chapter 2) as an example. The group is closed because composing any two displacements (translations, or shifts in space) results in another displacement. There is an identity displacement (not moving space at all), and for every displacement, there is an inverse displacement that undoes it (shifting the same distance in the opposite direction). Piaget described the fourth group property for displacements as "independence of path." Consider the example illustrated on the left side of in Figure 3.1.

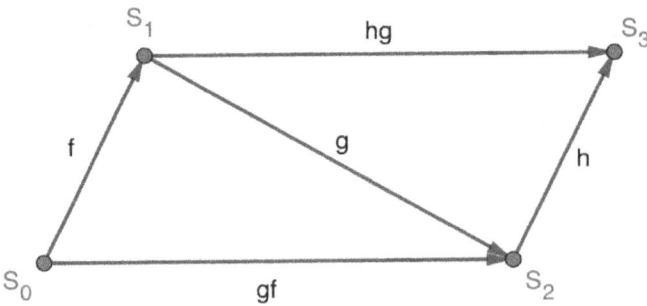

Figure 3.1 Associativity as independence of path.

This independence of path owes to the nature of spatial transformations. In the case of displacements, every point is space is shifted to another point in space so that there is a one-to-one correspondence between points before and after the displacement. If f, g, and h are displacements of a space, S, then f maps one version of S, S_0, to another version, S_1. Likewise, g maps S_1 to S_2, and h maps S_2 to S_3. gf maps S_0 to S_2 directly, by cutting out the middleman, S_1, and hg maps S_1 to S_3 directly, by cutting out S_2.[5] So, h(gf) is the same mapping as (hg)f. Both map S_0 to S_3.

Reflection

Knowing what we know about reflections and rotations from Chapter 2 and given the definition earlier, think about whether the set of all reflections in the plane form a group. What about the set of all rotations in the plane?

In Chapter 2, we extended Piaget's group of displacements to the group of isometries, all generated by reflections. In this chapter, we extend further to Klein's principal group for Euclidean geometry, and we consider how the transformations within that group define simple geometric figures.

CLASSIFYING QUADRILATERALS

To instantiate the ways that groups structure the mental actions we use to construct mathematical objects, we focus on a particular kind of geometric object: quadrilaterals. Quadrilaterals are closed, four-sided figures in the plane. Familiar examples include rectangles and trapezoids. Geometry courses usually define these special types of quadrilaterals based on relationships between their angles and sides. For example, we can define a *trapezoid* as a quadrilateral with exactly one pair of parallel sides. Figure 3.2 illustrates a common classification of quadrilaterals based on such properties. High school geometry classes sometimes require students to represent this classification as a "quadrilateral mobile," in which sets of shapes higher on the mobile contain the sets of shapes below (e.g., the set of rectangles and the set of rhombuses are contained in the set of parallelograms).

When we focus on geometry in terms of transformations of space, not only can we classify different kinds of geometries, but we can also classify geometric figures differently. For example, we can classify quadrilaterals based on transformations that leave them invariant. In other words, we can classify them based on their symmetries.

Recall from Chapter 2, we identified several transformations that preserve internal relationships between angles and sides of geometric figures. Chief among these isometries were reflections. When we consider quadrilaterals, squares have optimal reflective symmetry (see Figure 3.3).

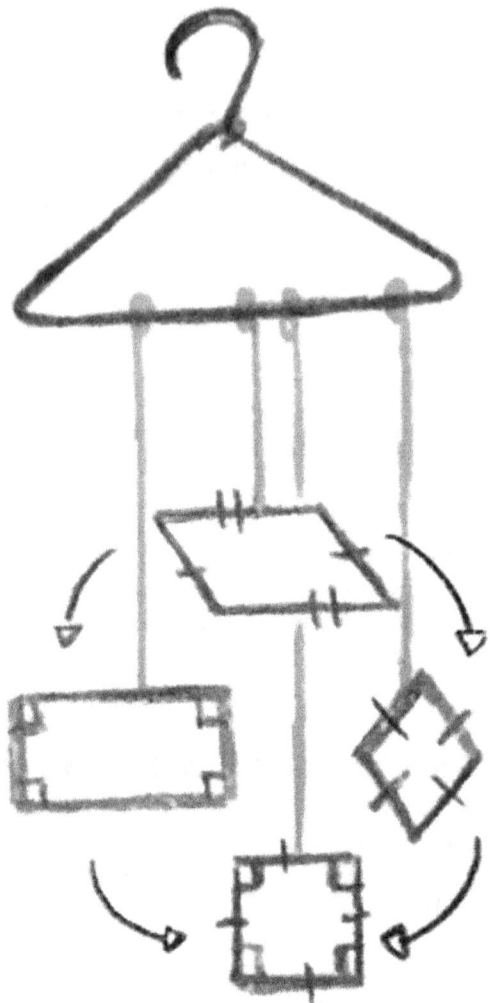

Figure 3.2 Mobile of quadrilaterals.

Source: © Eleanor Norton

The group of symmetries of the square includes four lines of reflection. Note that composing any two neighboring reflections results in a rotation of 90 degrees, indicating that the square also has 90-degree rotational symmetry. This is just a special case of composing reflections, discussed in Chapter 2. Composing the 90-rotations demonstrates that the square also has rotational symmetries of 180 and 270 degrees.

Now there are two primary subgroups of the symmetries of a square: one generated by the lines of reflection through the middle of opposite sides (the vertical and horizontal lines in Figure 3.3); and one generated by the diagonal lines of

Figure 3.3 Reflective symmetry of the square.

reflection (the diagonal lines in Figure 3.3). These subgroups define the symmetries of the rectangle and rhombus, respectively (see Figure 3.4). Note that each of them also includes a 180-degree rotational symmetry, as the composition of their pair of reflective symmetries.

These two subgroups share a common subgroup that has no reflective symmetry at all but does have a rotational symmetry of 180 degrees: the parallelogram. Moreover, each of the two subgroups has a subgroup of its own.

Reflection

What kind of quadrilateral would have a single line of symmetry through the middle of opposite sides, or a single line of symmetry through a diagonal? Half of each answer is shown in Figure 3.5. You can readily imagine the other half of the answer, much like you could for the tiger in Chapter 2.

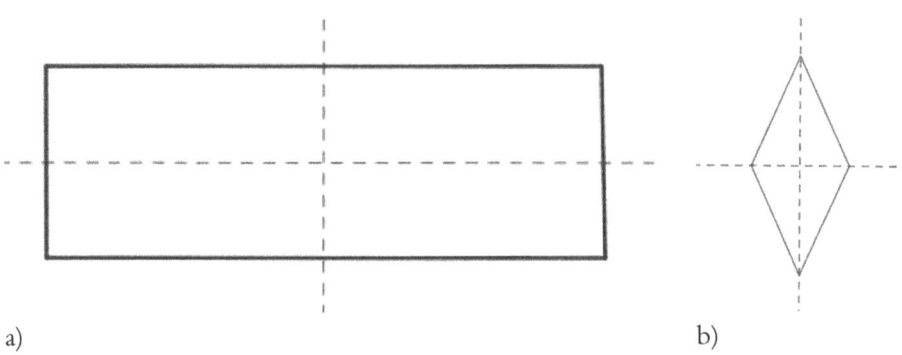

Figure 3.4 Reflective symmetries of the rectangle (left) and rhombus (right).

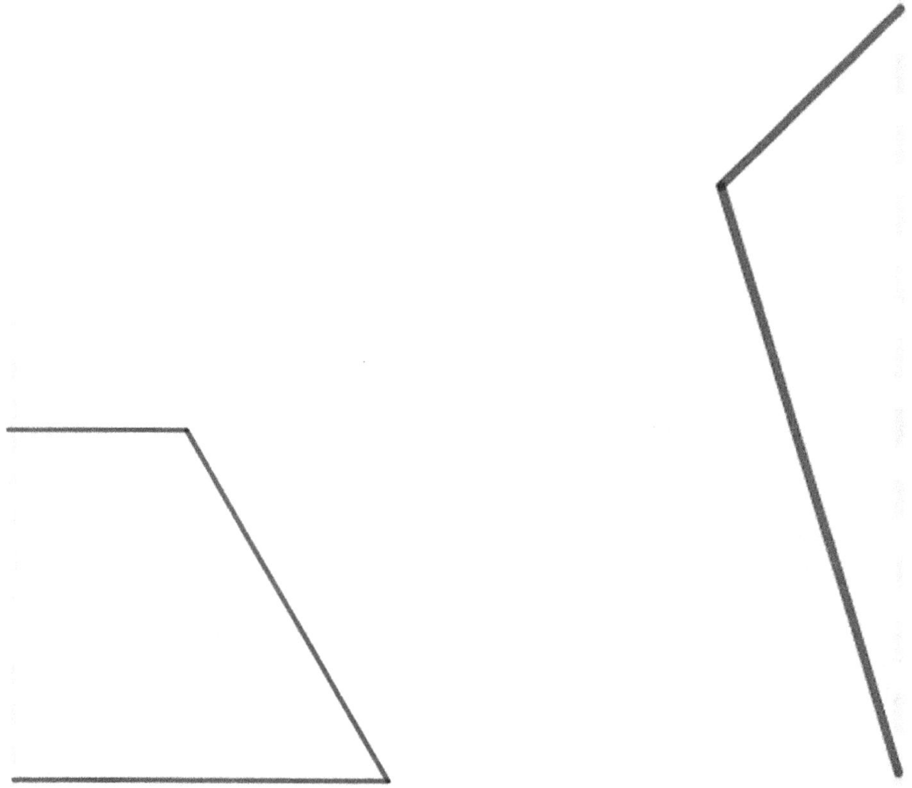

Figure 3.5 Half of an isosceles trapezoid (left) and half of a kite (right).

Think of our mobile now. Rather than the organization illustrated in Figure 3.2, we have the one suggested by Figure 3.6. With quadrilaterals defined by their symmetries, squares have the largest group: four lines of reflection, which also

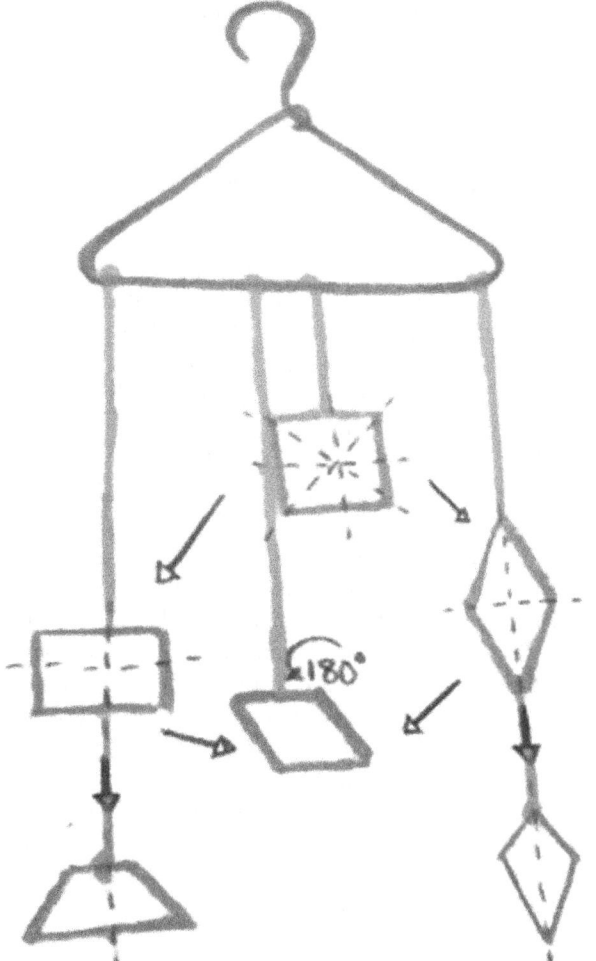

Figure 3.6 A transformational classification of quadrilaterals.

Source: © *Eleanor Norton*

generate rotations of 90, 180, and 270 degrees. One pair of those reflections generates the symmetries of the rectangle, and the other pair generates the symmetries of the rhombus. Those subgroups have in common the 180-degree rotational symmetry of parallelograms in general. Then, single reflective symmetries define the isosceles trapezoid and kite.

Defining quadrilaterals by their symmetries fits the Erlangen program of classifying geometries based on transformations that leave their objects (e.g., quadrilaterals) invariant, only now the program has been extended to define the geometric objects themselves. Because these transformations are reversible and composable mental actions, these definitions of quadrilaterals also follow Piaget's epistemology, explaining mathematical development as a product of our own psychology. Let's consider the mathematical and psychological power of this transformational approach to geometry.[6]

THE TRANSFORMATIONAL APPROACH

There are many ways to define geometric figures. Typically, rectangles are defined as four-sided, closed figures (quadrilaterals) with four right angles. They can also be defined as quadrilaterals whose diagonals are congruent and bisect each other. From a transformational perspective, figurative properties of the rectangle, such as angle measures and diagonal lengths, are secondary to its symmetries—two lines of reflections and a 180-degree rotation. We might just as well represent the rectangle by drawing its diagonals as by drawing its sides. Both figures represent quadrilaterals with the same two lines of symmetry (see Figure 3.7).

If we define the rectangle transformationally, based on its symmetries, the figurative properties of its sides, angles, and diagonals follow. In particular, the horizontal line of reflection shown in Figure 3.6 guarantees that the side and pair of angles at the top of the rectangle will be congruent to the side and pair of angles at the bottom of the rectangle (see the left side of Figure 3.7); it also guarantees that the top halves of the diagonals of the rectangle will be congruent to the bottom halves (see the right side of Figure 3.7). Likewise, the vertical line of reflection guarantees that the side and pair of angles on the left of the rectangle will be congruent to the side and pair of angles on the right of the rectangle and that the left halves of the diagonals of the rectangle will be congruent to the right halves. Together, these lines of symmetry guarantee that the rectangle will have four congruent angles and four congruent half-diagonals (i.e., the diagonals are congruent and bisect each other).

By relying on the transformations that define rectangles, we can derive powerful, intuitive arguments. For example, we can gain the insight Thales had about right angles 3,000 years ago: Any angle on a circle that is subtended by the diameter

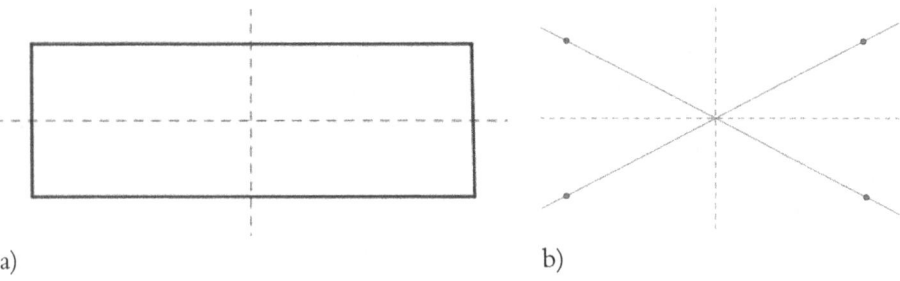

a)　　　　　　　　　　　　　　　　　　　　b)

Figure 3.7 Rectangles as dynamic objects.

of the circle must be a right angle. In other words, if the diameter of a circle forms one side of a triangle, and its opposite vertex is on the circle, the angle at that vertex must be 90 degrees (see Figure 3.8).

Sometimes lauded as "the first Greek mathematician," Thales lived around 600 BC.[7] He drew on early geometric ideas from the Babylonians and Egyptians to establish fundamental ideas for Greek geometry, including Euclid's *Elements*. Written around 300 BC, the *Elements* contained the first formal geometric proofs. Its proof of Thales' theorem appears as Proposition 31 in Book III.[8] It goes something like this:

Let O be the center of a circle with diameter AB, and draw line segment CO.
OA, OB, and OC are congruent because they are radii of the same circle.
The angle at A is congruent to angle Y because they are base angles of isosceles triangle AOC.
Likewise, the angle at B is congruent to angle X because they are base angles of isosceles triangle BOC.
The sum of the measures of the angles at A, B, and C is 180 degrees because they form the interior angles of triangle ABC.

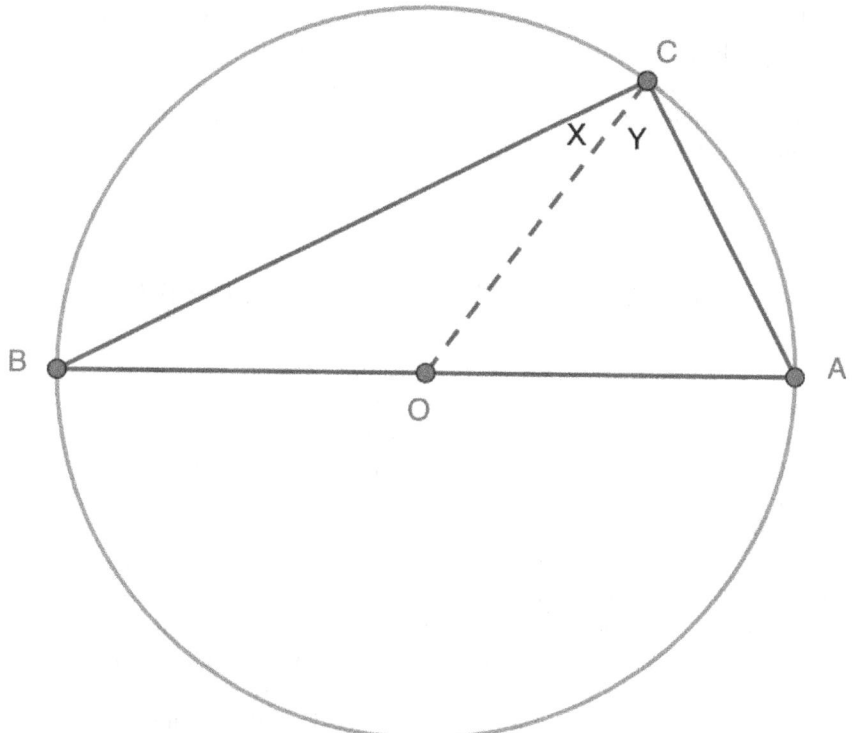

Figure 3.8 Thales' theorem.

But, recall that A = X, B = Y, and C = X + Y, so A + B + C = X + Y + (X + Y) = 180. Therefore, X + Y = 90.

Now we prove Thales' theorem intuitively by relying on the transformations that define rectangles. Although formal proofs contribute a kind of rigor to mathematical arguments, as demonstrated by Euclid (and reconsidered in Chapter 5), the transformational approach appeals more to the intuitions on which Thales must have relied. The transformational approach can make those intuitions more rigorous too by specifying relationships between the reversible and composable mental actions that define mathematical objects.

We have seen how rectangles can be defined by their symmetries: the transformations that leave them invariant in terms of their position in the plane. Those symmetries guarantee that the four angles in a rectangle must be right angles (see the left side of Figure 3.7). They also guarantee that the two diagonals of a rectangle must have the same length and must bisect each other (see the right side of Figure 3.7). In other words, any quadrilateral with the two lines of reflection illustrated in Figure 3.7 must have four congruent angles and have congruent diagonals that bisect each other.

Suppose we have a circle with two diameters drawn (see Figure 3.9). Those diameters intersect at the center of the circle and form the diagonals of a quadrilateral. Based on our transformational definition, we see immediately that the quadrilateral must be a rectangle. Specifically, the pair of diameters/diagonals have two lines of reflection as symmetries: the two lines of symmetry that define a rectangle. That means that the endpoints of the diameters (diagonals) will be vertices of a rectangle, where there are right angles, just as Thales claimed.[9]

GETTING WITH THE PROGRAM

So far, we have considered transformations that preserve geometric figures, thus defining geometric objects, such as rectangles, based on their reflective and rotational symmetries. When we include translations, these shape-preserving transformations form the group of isometries discussed in Chapter 2. If we allow transformations that preserve shape but change size, we can also include dilations, extending the group of isometries to the "principal group."

Dilations magnify or shrink geometric figures but maintain their angles and the relative lengths of their sides. Consider the dilation of a trapezoid, as illustrated in Figure 3.10. The dilation, from point O, scales up the smaller trapezoid by a factor of 2, producing a similar trapezoid where all the angles are preserved but each side has twice the original length. A reverse dilation would scale the larger trapezoid back down (by a factor of 1/2) to reproduce the smaller trapezoid. With the inclusion of such transformations, and their combinations with isometries

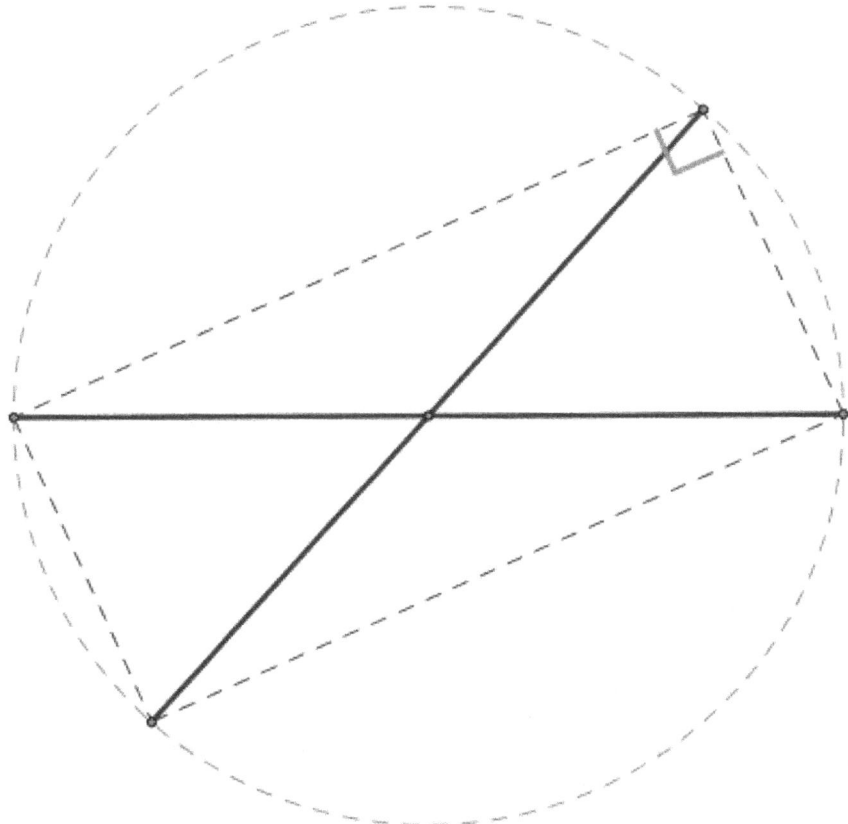

Figure 3.9 A transformational proof of Thales' theorem.

(e.g., reflections, rotations, and translations), we now have the principal group that Klein described in his Erlangen program.

Klein's principal group is the centerpiece of the Erlangen program. As mentioned at the start of this chapter, the Erlangen program classifies different geometries based on the groups of transformations that preserve their figures. The principal group preserves figures from Euclidean geometry (the geometry invented by Euclid, which is taught in high school). It is the group of mental actions that transform space but preserve geometric shape.

We have considered symmetries of quadrilaterals as subgroups of the principal group that define and classify those quadrilaterals. For example, the square is defined by a subgroup of the principal group that contains four lines of reflection and four rotations of 90, 180, 270, and 360 degrees. The four reflections and four rotations that compose the symmetries of a square might transpose its sides or angles, but they map the square, as a whole, onto itself.

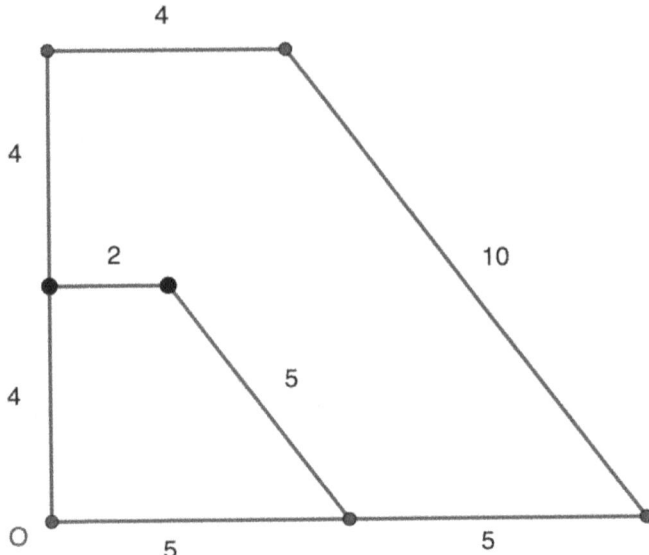

Figure 3.10 Dilation of a trapezoid.

In contrast, mental actions from the principal group will transform one square in space to another square somewhere in that space but not necessarily onto the self-same square with its particular size, position, and orientation in that space. In other words, whereas the principal group describes transformations that preserve geometric figures (like triangles and quadrilaterals), groups of symmetries describe transformations that preserve particular classes of those figures (like squares).

Now, we want to relate the principal group to the groups of symmetries that define particular classes of figures. We know these symmetry groups are subgroups of the principal group, meaning that every symmetry is one of the transformations from the principal group. We also know that every transformation from the principal group preserves geometric figures. The special property of the transformations in a figure's symmetry group is that they map the figure onto itself, leaving it fixed in the same position in space. Klein formulated the problem as follows:

> As the geometrical properties of configurations in space remain unaltered under all of the transformations of the principle group, it is by nature of the question absurd to inquire for such properties as would remain unaltered only a part of those transformations. This inquiry becomes justified, however, as soon as we investigate the configurations of space in their relation to elements regarded as fixed.[10]

Let's take the square as an example. What transformations from the principal group leave the square fixed as a whole? The answer is the symmetry group of the square. As Klein went on to argue, "it is exactly the same thing" whether we consider transformations with respect to the square (its symmetries) or whether, instead, we take those transformations from the principal group that leave the square fixed.

DUALITY AND PROJECTIVE GEOMETRY

Klein went even further in his arguments, noting how one configuration, such as a rectangle, might be transformed to another configuration, such as a rhombus, and how this transformation would also transform the symmetry groups that define those configurations. In particular, we might transform a rectangle into a rhombus by the dual transformation, which maps each side of the rectangle to a vertex and each vertex of the rectangle to a side (see Figure 3.11).[11]

This dual transformation maps any given rectangle to a corresponding rhombus and thus maps the symmetries of the rectangle to the symmetries of the rhombus. We can see how this would happen, the lines of symmetry on the left side of Figure 3.3 would become the lines of symmetry on the right side of Figure 3.3. In the special case of a square, which is both a rectangle and a rhombus, the dual would map one pair of lines of symmetry to the other pair and vice versa.

Reflection

What is the dual of a kite, and how does this relate to its symmetry group?

The dual transformation is not an element of the principal group. In fact, there are many transformations that we can consider beyond those found in the principal group. For example, we can expand the principal group to include projections, like the one illustrated in Figure 3.12.

Unlike transformations from the principal group, projections do not necessarily preserve geometric figures like squares and rectangles. Figure 3.12 shows a projection turning a square into a trapezoid. We also see this happen when video projectors distort rectangular images into trapezoidal images on the screen. Instead of preserving figures from Euclidean geometry, like squares and rectangles, projections preserve figures from projective geometry, where rectangles and trapezoids are considered to be the same. Extending the principle group to include projections generates a new group that characterizes projective geometry and, at the same time, classifies Euclidean geometry as a special case of projective geometry.

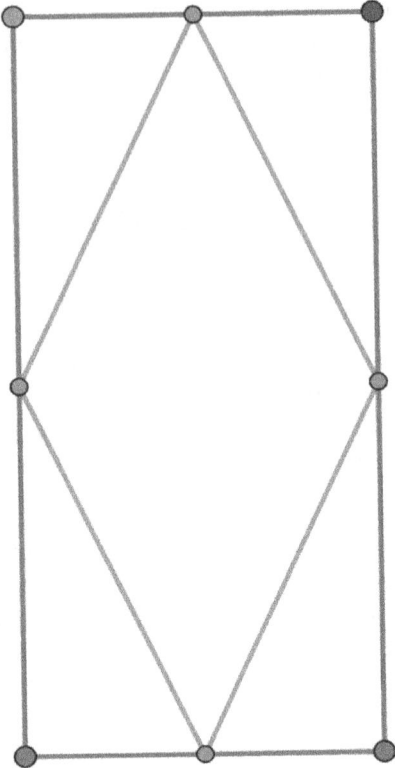

Figure 3.11 Dual mapping between a rectangle and a rhombus.

SUMMARY

Felix Klein developed the Erlangen program to classify various geometries (e.g., Euclidean geometry and projective geometry), but the program also closely aligns with Piaget's epistemology in three key ways. First, the group structure Klein relies on fits Piaget's characterization of mathematical mental actions as composable and reversible. Second, many of the transformations included in those groups are basic mental actions children learn to perform, such as reflecting (a primitive mental action) and translating (as in the group of displacements). Finally, the Erlangen program aptly shifts our attention from the figures we use to keep track of transformations to the mental actions that undergird them. We examined this shift closely in the case of quadrilaterals, which derive their properties from symmetries rather than side lengths.

Quadrilaterals provide just one class of figures for carrying out and extending the Erlangen program. We could just as well define triangles, polygons, and other figures on the basis of mental actions whose coordinations comprise

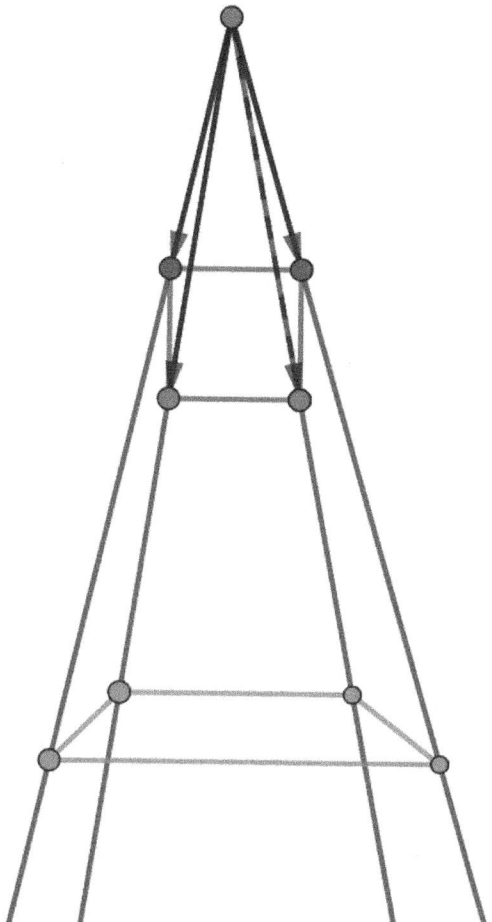

Figure 3.12 Projection of a square onto a trapezoid.

them or transformations that leave them invariant. We saw the psychological power of the approach in proving Thales theorem about right triangles inscribed in a circle. In Chapter 2, we saw another example, in proving that the interior angles in a triangle sum to 180 degrees. We can even extend the program beyond geometry.

In Chapter 4, we apply the Erlangen program to number. If we expand the whole numbers discussed in Chapter 1 to include fractions, we get a group that transforms the space of positive rational numbers. In Chapter 8, we expand numbers to include directions: positive and negative, left and right, or even a whole circle of directions. Such expansions admit groups that apply to larger spaces of numbers, such as complex numbers. Through these expansions, we will begin to see the integration of space and number.

Activities

Activity 1: Use the reflective symmetry of the isosceles triangle to argue that its base angles are congruent (see Figure 3.13).

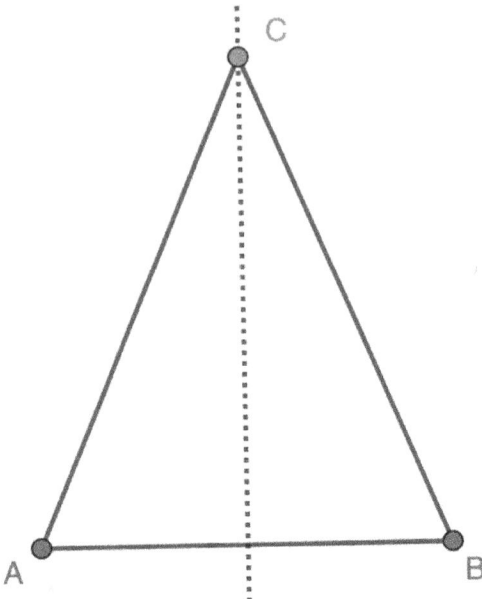

Figure 3.13 Isosceles triangle.

Activity 2: Use Figure 3.13 to argue why the perpendicular bisector of the base (side AB) must pass through the third vertex, C.

Activity 3: Match the following definitions with the appropriate quadrilateral.

Definition	Quadrilateral
• Diagonals bisect each other	• Isosceles trapezoid
• Diagonals are congruent	• Kite
• Diagonals meet at right angles where one diagonal is bisected	• Parallelogram
• Diagonals meet at right angles where both diagonals are bisected	• Rhombus

Activity 4: Match the following symmetries with the appropriate quadrilateral.

Symmetry	*Quadrilateral*
• One line of reflective symmetry	• Kite (or isosceles trapezoid)
• Two lines of reflective symmetry and 180-degree rotational symmetry	• Square
• 180-degree rotational symmetry	• Parallelogram
• Four lines of symmetry and 90-degree rotational symmetry	• Rectangle (or rhombus)

NOTES

1. Piaget's epistemology was especially influenced by Kant's (1998/1781) *Critique of Pure Reason.*

2. In a letter to Sophus Lie, Felix Klein explained, "I aim toward a complete sensory intuition of the things in space and the laws that hold them together" (letter written February 16, 1872, as translated by Leslie Kay).

3. "Sensori-motor space begins to evolve right from the child's birth, and together with perception and motor activity it undergoes considerable development up until the appearance of speech and symbolic images" (Piaget & Inhelder, 1967, p. 3).

4. In fact, the composition of functions, relations, and mappings will always be associative. In particular, if f, g, and h map a space, S, to itself, then $(gf)(S) = g(f(S))$, by definition of composition. Thus, $h(gf)(S) = (h(g(f(S))) = (hg)f(S)$ and $h(gf)h = (hg)f$, as required.

5. Note that, by convention, we read and write the composition of transformations from right to left. So, gf represents f followed by g. This is because transformations are functions, and in function notation, writing gf(x) would mean f acts on x first, and then g acts on the result.

6. Martin Simon's (1996) transformational reasoning describes a third form of reasoning—in addition to inductive and deductive reasoning—that is "the enactment of an operation or set of operations on an object or set of objects that allows one to envision the transformations that these operations undergo" (p. 201). As such, this form of reasoning aligns closely with Peirce's abductive reasoning and, especially, Norton's (2008) characterization of operational conjectures.

7. This quote comes from Fletcher (1982). Like many mathematicians throughout history, Thales was also a philosopher. He challenged us with the wisdom to "know thyself."

8. For an excellent presentation and illustration of Euclid's *Elements*, visit https://mathcs.clarku.edu/~djoyce/java/elements/elements.html

9. In a *Mathematician's Lament*, Paul Lockhart (2009) shared a similar, intuitive argument generated by one of his high school students. Referring to a triangle like the one shown in Figure 3.8, the student imagined rotating the triangle around to form a rectangle.

10. These quotes come from the speech Klein prepared to address faculty at the University of Erlangen upon joining their ranks in 1872. The speech was later translated and published in the *Bulletin* (1893, p. 219).

11. This dual transformation is its own inverse and forms an isomorphism between the symmetry group of the rectangle and the symmetry group of the rhombus.

REFERENCES

Fletcher, C. R. (1982). Thales—our founder? *The Mathematical Gazette, 66*(438).

Kant, I. (1998). *Critique of pure reason* (P. Guyer & A. Wood, Trans.). Cambridge: Cambridge University Press (Original work published in 1781).

Klein, F. (1872). *Unpublished letter written to Sophus Lie on February 16, 1872* (L. Kay, Trans.).

Klein, F. (1893). A comparative review of recent researches in geometry (M. W. Haskell, Trans.). *Bulletin of the American Mathematical Society, 2*(10), 215–249.

Lockhart, P. (2009). *A mathematician's lament.* New York: Belleview Literary Press.

Norton, A. (2008). Josh's operational conjectures: Abductions of a splitting operation and the construction of new fractional schemes. *Journal for Research in Mathematics Education, 39*(4).

Piaget, J., & Inhelder, B. (1967). *The child's conception of space* (F. J. Langdon & J. L. Lunzer, Trans.). New York: Norton (Original work published in 1948).

Simon, M. A. (1996). Beyond inductive and deductive reasoning: The search for a sense of knowing. *Educational Studies in Mathematics, 30*(2).

4

New Actions, New Units, New Numbers

W hen we, as adults, see a numeral, like 7, we can take its meaning for granted, confident that others understand it the way we do, as seven 1s. Chapter 1 outlines the laborious process that leads to this conception of whole numbers, through childhood and across cultures. We take it for granted because that development, however challenging, occurs so early in our lives and has become foundational to the ways we organize the worlds we perceive, as in seeing seven yellow bricks. But as we consider new units, other than 1, there is new work to be done.

Recall that composite units are units made up of other units (Chapter 1). For example, we might treat 7 as a composite unit—a unit made up of seven 1s—and consider the value of five 7s. To determine that value, we need to consider two levels of units: the five units of 7 as well as the seven units of 1. This coordination across two levels of units is the basis for multiplicative reasoning. Now, we will consider three new kinds of units: unit fractions, decimals, and unknowns.

FRACTIONS

When we think of a fraction, like 5/7, it might take on different meanings for us. For example, it might represent 5 divided by 7 or a 5-to-7 ratio. Typically, we learn

DOI: 10.4324/9781003181729-5

Figure 4.1 Diagram of 5/7 as five parts out of seven parts in the whole.

in school that it represents five parts out of seven equal parts in a whole, and we might use a drawing like the one shown in Figure 4.1 to illustrate that meaning.

Part–whole conceptions of fractions, such as 5 out of 7, pervade school curricula, particularly in the United States.[1] There is nothing wrong with this conception, but it is inherently limited, especially when we begin to consider improper fractions, like 9/7. How can you take 9 out of 7? This is one of the reasons students are uncomfortable with fractions and the main reason they try to avoid improper fractions, converting them to mixed numbers (e.g., 1 2/7) instead.

To fully understand fractions, a reliance on whole units is insufficient. Fractions are more than a comparison of two whole numbers, like 5 and 7. We cannot conceptualize improper fractions, fraction multiplication, or other operations on fractions by simply comparing the number of parts in the fraction to the number of parts in the whole. We need a new kind of unit for measuring—the unit fraction as a fractional unit.

UNIT FRACTIONS

We make unit fractions by breaking a whole into equal parts and taking one of them, such as 1/5 or 1/7, as a new unit. Here, we consider unit fractions as a multiplicative relationship between the part and the whole. For example, seven units of 1/7 make up a whole unit. As mathematical objects, unit fractions arise from a coordination of mental actions. In this case, the principal mental actions are *partitioning* and *iterating*.

Like counting, the mental action of partitioning develops in stages, beginning with the coordination of sensorimotor activity. Young children might engage in sharing activities by breaking up collections or continuous wholes. For example, they might want to share a candy bar fairly among five friends. In doing so, there are three competing goals at play:

1. There should be five shares.
2. Each share should be the same size.
3. There should be nothing left over.

Satisfying all three goals at once requires experience in coordinating sensorimotor experiences of counting, fragmenting (breaking up a continuous whole into parts), and comparing sizes, possibly by lining them up. Oftentimes, children will satisfy two of the goals, ignoring the third goal: they might create five equal parts but have a leftover part that they do not consider, or they might break the bar into five unequal parts, or they might focus on creating equal parts but make too many because the part they started with was too small.

Once children can coordinate their activity to satisfy all three goals at once, we attribute them with the mental action of partitioning. Like the mental actions we have discussed in geometric contexts, such as reflecting and rotating, partitioning is a reversible and composable mental action that empowers our construction of mathematical objects, like fractions.

The inverse action for partitioning is iterating, the mental action of making connected copies of a unit. Consider the example illustrated in Figure 4.2. If a whole has been partitioned into five equal parts, we can take any one of those parts as 1/5 and iterate it five times to reproduce the whole. This reversible relationship between the whole and 1/5 establishes 1/5 as a 1-to-5 relationship, a unit fraction multiplicatively related to the whole.

The 1-to-5 relationship allows us to measure other fractions, like 3/5, in units of 1/5. Figure 4.2 illustrates the situation, with a whole bar (top), partitioned into five equal parts with one fifth disembedded (middle), and then that part iterated three times to produce 3/5 (bottom). As we will see, this relationship—established by a coordination of our own mental actions—frees us from part–whole conceptions and enables us to conceptualize all fractions, even improper fractions, as "numbers in their own right."[2] It also enables us to meaningfully operate on fractions, as in fraction multiplication and fraction division. In other words, it allows us to treat fractions as a system of numbers, described in the next section as the splitting group.

Figure 4.2 1/5 as a 1-to-5 relationship with the whole and 3/5 as three iterations of 1/5.

THE SPLITTING GROUP

In Chapter 3, we considered groups of mental actions that undergird concepts in geometry. Here, we consider a new group, one that undergirds conceptions of fractions. Just as the group of isometries is generated by the mental action of reflecting, the splitting group is generated by the mental actions of partitioning and iterating.[3] The group structure describes how coordinating those mental actions, especially their various compositions, generates concepts for all fractions—unit fractions, proper fractions, and improper fractions, alike.

Table 4.1 displays the structure of the splitting group. P_n symbolizes the mental action of partitioning a whole into n equal parts. I_m symbolizes the mental action of iterating a part m times. The entries in the table show the result of composing various combinations of partitioning and iterating. For example, the entry for the fourth row and fifth column is the composition of I_5 with P_4, which produces the fraction 5/4. In other words, if you partition a whole into four equal parts, you produce the unit fraction, 1/4, which when iterated five times, produces 5/4.

Of course, the table would continue indefinitely, down and to the right, to include ever larger numbers for partitioning and iterating. Note that it also includes whole numbers (in the top row) and the possibility for equating fractions (e.g., 2/4 = 1/2). However, determining the equivalence of two fractions requires additional coordination. In the example of 2/4 and 1/2, we would need to determine that two iterations of a 1/4 unit measure are the same amount as one iteration of a 1/2 unit.[4]

Recall that groups are closed under composition. Thus, the splitting group must contain all the fractions in Table 4.1, in addition to the partitions and iterations (e.g., P_4 and I_5) that form them. We can see that partitioning and iterating generate this group, but the group itself includes much more. It is the group of all fractions under the operation of multiplication, and we will use this group to demonstrate how fraction multiplication and division arise from a coordination of the basic mental actions of partitioning and iterating.

Table 4.2 displays a small piece of the complete group. Note that the entries in the top row and leftmost column are already coordinations of partitioning and iterating, as indicated in Table 4.1. For example, 2/3 is the composition of P_3 with I_2, and 3/2 is the composition of P_2 with I_3. So, multiplying 2/3 by 3/2 involves the composition of four mental actions: P_3, I_2, I_3, and P_2.

Table 4.1 Generating the splitting group

	I_1	I_2	I_3	I_4	I_5
P_1	1	2	3	4	5
P_2	1/2	1	3/2	4/2	5/2
P_3	1/3	2/3	1	4/3	5/3
P_4	1/4	2/4	3/4	1	5/4
P_5	1/5	2/5	3/5	4/5	1

Table 4.2 A snippet of the splitting group

	1	1/2	1/3	2/3	3/2
1	1	1/2	1/3	2/3	3/2
1/2	1/2	1/4	1/6	2/6	3/4
1/3	1/3	1/6	1/9	2/9	3/6
2/3	2/3	2/6	2/9	4/9	1
3/2	3/2	3/4	3/6	1	9/4

Reflection

Using the definition of a group given in Chapter 3, can you verify that the splitting group does indeed form a group? In particular, if 1 is the identity, can you verify that every fraction has an inverse and that the composition of any two fractions forms another fraction?

FRACTION MULTIPLICATION

We saw in Chapter 1 that multiplication involves a transformation of units. For example, 4 × 3 is the transformation of four units of 3 into twelve units of 1 (see Figure 1.7 in Chapter 1). This transformation is enabled by the composite unit of 3, which is simultaneously understood as three units of 1: three units of 1 get inserted into each of the four units of 3. It is the same with fractions.

When we multiply 2/3 × 3/2 we are transforming units, except that 2/3 and 3/2 are already the result of transforming units. Consider the diagrams in Figure 4.3. 2/3 is a transformation from units of 1/3 to units of 1; namely, 2/3 is two units of 1/3 and becomes 2/3 of 1. Likewise, 3/2 is a transformation from units of 1/2 to units of 1; it is three units of 1/2 and becomes 3/2 of 1. To appreciate the additional complexity of fraction multiplication over whole number multiplication consider the following analogy: 2/3 is the answer to the question, "How much is two units of 1/3 when measured in units of 1," just as 12 is the answer to the question, "How much is 4 units of 3 when measured in units of 1?" With fractions multiplication, we are multiplying two numbers, where each number is already as complex as whole number multiplication.

Just as geometric transformations rely on the coordination of mental actions, so do transformations of units. The mental action, P_3, transforms the whole unit into a 1/3 unit, which we can iterate twice (I_2) to produce 2/3. Likewise, 3/2 is the unit transformation resulting from P_2 followed by I_3. Because multiplication is itself a transformation of units, the product 2/3 × 3/2 is just a composition of the four underlying mental actions: $P_3 I_2 P_2 I_3$. What is the result of that composition?

We saw in Chapter 3 that composing transformations of space is always associative because one transformation picks up where the prior one left off. The same

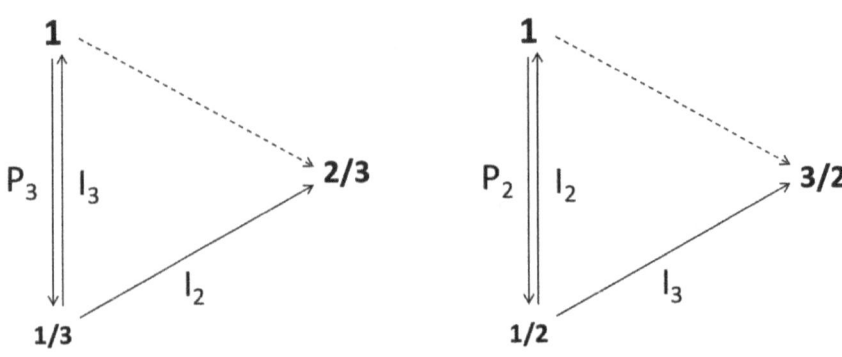

Figure 4.3 2/3 and 3/2 as transformations of units.

is true for composing mental actions like partitioning and iterating, which transform (and indeed produce) a space of numbers that we call fractions. Associativity of partitioning and iterating, in particular, tells us we can combine the middle two actions first: $P_3 I_2 P_2 I_3 = P_3 (I_2 P_2) I_3$. But iterating and partitioning are inverse actions, so $(I_2 P_2)$ is just the identity, which in the case of multiplication is just 1. We are left with $P_3 I_3$, which (for the same reason) is just 1. Thus, we find that the two fractions are multiplicative inverses of each other because their composition results in the multiplicative identity, 1. In other words, the mental actions undo each other, yielding the whole with which we started. We can easily generalize the argument to see why the reciprocal of a fraction (switching its numerator and denominator) will always yield its multiplicative inverse. Namely, the reciprocal of m/n is n/m, and their product is the composition, $P_n I_m P_m I_n$, which yields 1.

COMMUTATIVITY OF MULTIPLICATION

We have argued that all mental actions that transform a space—whether it be a geometric space or a space of numbers—are associative under composition. However, we can't take for granted that mental actions will commute, and unlike associativity, commutativity is not an axiom of groups. For example, if we consider the pairs of reflections from Figure 2.2 in Chapter 2, composing the mental actions in one order will yield a shift to the right, or a counterclockwise rotation, and composing them in the other order will yield a shift to the left, or a clockwise rotation. So how can we determine whether partitioning and iterating commute?

Children typically don't understand that multiplication is commutative, even for whole numbers. They have to learn this through their own activity. For example, a child might see 5 × 8 and 8 × 5 as two different multiplication problems. They are likely to see 5 × 8 as harder because five 8s are not so easy to coordinate as eight 5s (eight 5s is just four 10s because every pair of 5s makes a 10). Children who understand multiplication as commutative can take advantage of this property to convert the harder problem into the easier one because, after all, they are the same. Likewise, understanding the commutativity of fraction multiplication can be useful in converting problems like finding 2/5 of 8 into the problem of finding 8 of 2/5, which is just 16/5.

Note that the commutativity of I_m and I_n, depends on the commutativity of whole number multiplication (i.e., $m \times n = n \times m$). In Chapter 1, we characterized multiplication as a transformation of units. So, if we want to show that whole number multiplication is commutative, we need to show that the two transformations shown in Figure 4.4 produce the same result.

Our formal argument will follow Euclid's original proof of the claim.[5] On the right side of Figure 4.4, $n \times m$ is measured in units of m: $n \times m$ is n units of m. However, n itself is n units of 1. So, $n \times m$ is measured in units of m the same number of times n is measured in units of 1 (i.e., 1 is to n as m is to $n \times m$). By repeating the 1-to-n ratio m times, we get an equivalent m-to-m × n ratio, which tells us that $m \times n$ is measured in units of m the same number of times n is measured in units of 1. Therefore, $m \times n$ and $n \times m$ are the same measure—both are n units of m.

As exemplified in the preceding argument, numerical reasoning requires us to coordinate various levels of units. Each of the two triangles in Figure 4.4 represents a three-level unit structure: there are units of 1; units of n (or m), where each unit of n is n units of 1; and a unit of $m \times n$ (or $n \times m$), which is a unit of m units of n 1s. It's a lot to maintain, but this is the mental gymnastics children develop through mental arithmetic. Once children develop such three-level unit structures, they can understand numbers like 24 as a unit of eight units of 3, each of which is three units of 1. They can flexibly work within such structures to reconstitute 24 as three units of 8, each of which is eight units of 1. Children's understanding of commutativity of multiplication, then, owes to flexible reasoning within three-level unit structures. In fact, reasoning with units structures has far-reaching implications for children's learning, from whole-number addition to fractions division to algebraic reasoning.[6]

As we have noted, constructing fractions and performing fractions multiplication come down to a coordination of partitioning and iterating. Thus,

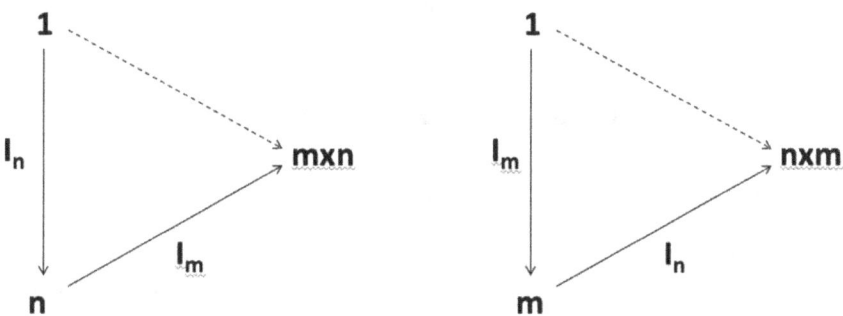

Figure 4.4 Commutativity of multiplication.

Figure 4.5 Commutativity of partitioning and iterating.

demonstrating the commutativity of fraction multiplication comes down to demonstrating the commutativity of those two mental actions. We need to show that partitioning the iteration of a unit is the same as iterating the partitioning of a unit. Let's say, for example, we have some unit length (it might be a whole unit, a unit fraction, or any other measurable length) and we iterate it five times; then we partition it into eight equal parts and consider the resulting size of one of those parts (see the four bars at top of Figure 4.5). Will it be the same if we, instead, partition the original unit into eight equal parts and then iterate one of those parts five times (see three bars at bottom of Figure 4.5)? Is one-eighth of five the same as five of one-eighth?

The answer is yes, which we can explain as a consequence of the splitting group, wherein partitioning and iterating act as inverses of each other. Specifically, the compositions P_nI_n and I_mP_m result in the identity element, 1. So, $(P_nI_n)(I_mP_m)$ is also 1. Because composition is associative, this is the same as $P_n(I_nI_m)P_m$. Moreover, $I_nI_m = I_mI_n$ because both are just $I_{m\times n}$ (see Figure 4.5). Now we have $P_n(I_mI_n)P_m = 1$. Relying on associativity once more, we have that $(P_nI_m)(I_nP_m) = 1$. In other words, (P_nI_m) and (I_nP_m) are inverses of each other, but recall that the inverse of (P_nI_m) is P_mI_n. So, $I_nP_m = P_mI_n$, which means iterating and partitioning commute.

FRACTION DIVISION

The product of two fractions m/n and s/t will be ms/nt. We can see this in the composition of underlying mental actions, $P_nI_mP_tI_s$. By commutativity of these mental actions, this composition is the same as $P_nP_tI_mI_s$, which (by associativity) is $P_{nt}I_{ms}$, as expected.[7] Now, we can unravel one of the more mysterious fraction rules—the keep-change-flip rule for dividing fractions—by similarly appealing to the underlying composition of mental actions.

In school, students often learn that to divide two fractions, you keep the first fraction the same, change the operation from division to multiplication, and flip (or take the reciprocal of) the second fraction. Although the rule works, it offers little understanding of fraction division and, understandably, few students are able to explain it or even give a practical example of when they would use fraction division. Blindly following the rule provides no more insight than simply punching in the numbers on a calculator. So, why does the keep–change–flip rule work?

Consider the example of 3/4 divided by 2/3. The keep–change–flip rule tells us the answer will be 9/8, but why? Formally, the division of one fraction by another is just multiplication of the first fraction by the multiplicative inverse of the second fraction (invert and multiply), and as we have noted, the multiplicative inverse of a fraction is its reciprocal. Thus, 3/4 divided by 2/3 would be the same as 3/4 times 3/2, which is exactly what the keep–change–flip rule tells us. However, we want to operate at the level of the mental actions that undergird formal mathematics and rules alike.

If multiplication is a transformation of units, division reverses the transformation. 9/8 times 2/3 gives the value of 9/8 of a 2/3 unit, measured in units of 1, which is 3/4. Figure 4.6 illustrates the situation by showing a whole bar (top), partitioned into three equal parts, with two-thirds disembedded. Then, that two-thirds bar is unitized as a new whole that is partitioned into eight equal parts, with one of those parts iterated nine times to produce 9/8 of the 2/3 bar. If we also partition the remaining 1/3 of the original (top) bar into eight parts, we see that this 9/8 of 2/3 is 9/12 (also known as 3/4) of the original bar.

In contrast, 3/4 divided by 2/3 answers the question of how many 2/3 units fit into 3/4, which is 9/8. In other words, if we measure ¾ in units of 2/3, we get

Figure 4.6 Fraction division.

9/8. Returning to Figure 4.6, we can treat the bottom bar as 3/4 of the top bar. Then we see that 2/3 of the top bar, when partitioned into eight equal parts, is 8/9 of the 3/4 bar. Conversely, the 3/4 bar is 9/8 of the 2/3 bar. Once again, we have to contend with a lot of units to figure this out, but ultimately, these units are nothing more (or less) than coordinations of our own mental actions—partitions and iterations, in particular.

Let's start with unit fractions. Suppose we want to know how many 1/3 units fit into a whole, or equivalently, we want to measure the whole in units of 1/3. This is not so hard. The result of partitioning the whole into three equal parts is 1/3, and the inverse action, iterating that part three times, will reproduce the whole. Okay, so how many 1/3 units fit into a 1/4 unit? The answer will be 1/4 as many, 1/4 of 3, or 3/4. Ultimately, we want to know the number of 2/3 in 3/4, so let's next consider the number of thirds in three fourths. This will be three times as many thirds as fit into one fourth, so we now have 9/4. Finally, to determine the number of 2/3 in 3/4, we take half of 9/4 because half as many 2/3 will fit, and we end up with 9/8.

To generalize this result and make the connection to mental actions more explicit, consider m/n divided by s/t, the number of times s/t fits into m/n. In terms of mental actions, m/n is the result of applying $P_n I_m$ to a whole unit (partitioning the whole unit into n equal parts and iterating one of those parts m times). Likewise, s/t is the result of applying $P_t I_s$ to a whole unit. Following the same progression as above, we know that P_t applied to a whole unit will result in 1/t and that t of those units will fit into the whole because, as the inverse of P_t, I_t applied to 1/t will reproduce the whole.

So far, we have done nothing but coordinate P_t with its inverse I_t to return to the whole unit from which we started. However, if we wanted to produce 1/n instead (P_n applied to the whole), we would have $I_t P_n$ (the number of times 1/t fits into 1/n). Furthermore, if we wanted to produce m/n ($P_n I_m$ applied to the whole), we would have $I_t P_n I_m$ (the number of times 1/t fits into m/n). Finally, if we wanted to produce m/n ($P_n I_m$ applied to the whole) from s/t ($P_t I_s$ applied to the whole) rather than just 1/t (P_t applied to the whole), we apply P_s to $I_t P_n I_m$ to account for the s iterations of P_t. The final result, $I_t P_n I_m P_s = I_m I_t P_n P_s = P_{ns} I_{mt}$ (by commutativity and associativity) aligns precisely with the keep–change–flip rule: m/n divided by s/t is mt/ns.

Note that, in performing these mental actions, it is useful to consider them applied to some figure, like a stick or a bar. However, the Erlangen program and Piaget's epistemology of mathematics challenge us to accept that mental actions themselves constitute mathematics and mathematical objects. We can represent fractions, like 2/3, in various ways, but the mental coordination of $I_2 P_3$ is their essence. In mathematics in general, we apply our mental actions to a space, such as Euclidean space or the space of fractions, but to see mathematics in its purest form, we need to let go of that space. When we do that, we realize there is nothing left but—and mathematics requires nothing more than—our own mental actions and our individual efforts to coordinate them in ever more complex ways.

OTHER REPRESENTATIONS OF NUMBER

So far, we have considered units of 1 (whole units), composite units, and unit fractions, and we have begun to coordinate various levels of units involved in numerical reasoning. Here, we consider how the decimal system provides us with a useful way of representing such units and operating on them (coordinating and transforming them). To appreciate the number system and why it presents a challenge for students to learn, we compare it to the binary system.

The Decimal System

We can think about decimal units as privileged fractions. For example, 0.1 is just 1/10, and this fraction is privileged because we happen to have 10 fingers on our hands. This is the very reason we use a base-10 system (and the reason we call the Arabic numerals, from 0–9, "digits"). If we had only four fingers on each hand, no doubt we would use a base-8 number system, and fractions like 1/8 and 1/64 (because 1/64 is $1/8^2$, just as 1/100 is $1/10^2$) would be privileged fractions. In this sense, our base-10 system is a product of our culture, as humans. It has appeared in virtually every human culture in the history of mathematics, across the world.

On four different continents, ancient Egyptian, Babylonian, Chinese, and Incan (Peruvian) cultures independently developed number systems, and all of them relied on grouping units of 1 into 10s. The Egyptians symbolized 1s with tally marks and symbolized 10 with something like a horseshoe, as if to bundle up ten units of 1 (see the left side of Figure 4.7). Likewise, the Babylonians tallied 1s on clay tablets with a stylus and used the triangular top of the stylus to make an impression for 10. However, unlike the Egyptians, who needed a new symbol for every new power of 10 (e.g., 100 and 1,000 had their own symbols), the Babylonians, Chinese, and Incas each independently invented a new idea that we rely on today in our own (Hindu-Arabic) base-10 system: place value.[8]

Reflection

What number is represented (in ancient Egyptian) on the right side of Figure 4.7?

In Chapter 1, we made a distinction between addition, which preserves units, and multiplication, which transforms units. That distinction is critical to understanding the way we write numbers within our base-10 system, and we can see roots of the idea in the Egyptian number system. Namely, the three symbols on the left side of Figure 4.7 represent three different levels of units: ones, tens as 10 ones, and hundreds as 10 tens or 100 ones. The Egyptians would reason additively within each level of units, simply iterating and marking a symbol for each kind of unit (like tally marks). For example, the right side of Figure 4.7 shows seven ones, four tens,

One   Ten   Hundred

Figure 4.7 Ancient Egyptian numerals.

and five hundreds. If the Egyptians wanted to add two numbers they would, again, preserve these units by adding each kind of unit separately. However, if the number of ones exceeded ten, they would group ten ones to make another ten. Likewise, they would group ten tens to make another hundred (similar to what we do when we "carry the 1"). In other words, they would transform the units.

The relationships between ones, tens, and hundreds imply a multiplicative transformation of units: to transform from units of one to units of ten, we can multiply by 10; to transform from tens to hundreds, again, we multiply by 10. This is also how we operate in our base-10 system. We operate additively within each level of units and then transform between levels of units by multiplying or dividing by 10.

Note that, in our Hindu–Arabic base-10 system, we do not have a symbol for ten or a hundred. We write these numbers, and all powers of 10, with a 1 followed by some number of 0s. This is because we rely on the positions of our numerals, 0 through 9, to indicate whether we are referring to ones, tens, hundreds, or any other power of 10; there is a ones place, a tens place, and so forth. We can also work in the other direction, for positions to the right of the ones place. If a position to the left should be 10 times more, a position to the right should be 10 times less, so the position to the right of the ones place should be the one-tenths place, as it is.

As decimal (base-10) numbers, tenths, hundredths, and thousandths are all privileged fractions. Moreover, our base-10 system provides us with an easy way of relating them: multiplying any of them by 10 moves them one place to the left; dividing any of them by 10 moves them one place to the right. For example, we know there are ten (10) one-thousandths (0.001) in a hundredth (0.01), and one hundred (100) thousandths (0.001) in one tenth (0.1). This structure for transforming units, in units of 10, is built into the way we represent numbers in the base-10 system. It becomes the basis for the rules we use to multiply and add decimals. If we used a different system, the rules would be different.

Here, we are making a distinction between numbers themselves and the numerical systems we use to represent them. Whereas numbers are products of our own coordinated mental actions, numerals and number systems are culturally defined ways of representing those numbers. True, there is a lot of commonality to the number systems different cultures have developed (owing to our ten fingers), but we can imagine them being otherwise. In fact, even although the Babylonians grouped ones into tens, their system was base-60, and the Mayans' system was base-20.[9]

The Binary System

To fully appreciate the complexity and utility of the decimal system, it may be helpful to consider a less familiar system: the base-2 (binary) system. In that system, rather than privileging tenths and hundredths, we privilege halves and fourths. Beginning from 1, we multiply by 2 for every position we move to the left, and we divide by 2 for every position we move to the right. For example, beginning from 1, 100 would represent four because we have moved the 1 two positions to the left (multiplying by 2 twice), and 0.001 would represent one eighth because we have moved the 1 to the right three times (dividing by 2 thrice).

We need only 1s and 0s to represent each number, because 2 has its own position. Two takes the place of ten so that 10 represents two in binary (1 two and 0 ones), just as 10 represents ten in our decimal system (1 ten and 0 ones). When you consider the complexity of learning a new positional system, like the base-2 system, you might better appreciate the challenges students must face in learning the base-10 system. Tasks like the following make the challenge clearer.

Reflection

What do you get when you add one, two, four, eight, sixteen, and thirty-two?

An analogous task in base-10 would probably seem silly to you: "What do you get when you add one, ten, one hundred, one thousand, ten thousand, and one-hundred thousand?" The answer would be represented the same way within each base system: 111,111. However, you might have difficulty interpreting what 111,111 represents in base-2. Well, it represents exactly what the task asks: a one, a two, a four, an eight, a sixteen, and a thirty-two. But how big is this number? To understand how much we rely on the base-10 system, it might be helpful to consider a different analogous task.

"What do you get when you add 9 ones, 9 tens, 9 hundreds, 9 thousands, 9 ten thousands, and 9 one-hundred thousands?" Of course, the answer is 999,999, but how big is that number? It's about a million. Specifically, it's one less than a million. We know that adding one more one will make another ten, which will make another hundred, and so on, until we have raised the levels of units to a new unit—a new power of 10, called one million. For 111,111 in base-2, it's the same situation. One more one will make a two, and one more two will make a four, and so on, until we have the next power of 2: sixty-four (64 in base-10). In base-2, this power of 2 has its own position, which is represented as 1,000,000. So, in base-2, 111,111 is 1,000,000 minus 1, just as, in base-10, 999,999 is 1,000,000 minus 1.

As with the base-10 system, positions in the binary system represent transformations of units, except these transformations are the result of multiplying or dividing by two rather than ten. If the binary system is qualitatively different than our base-10 system, it's because there is only room for one of each kind of unit (rather than up to nine iterations of units at each level within the base-10 system).

That renders the base-2 system ideal for representing products. For example, 10 times 10 is 100, both in base-10 and in base-2; but in base-2, we can readily relate all multiplication tasks to products like this. The Egyptians took advantage of this feature implicitly when using doubling and halving to multiply and divide numbers, as we will see in Chapter 6. We will take advantage of this feature of the base-2 system, explicitly, in Chapter 10.

UNKNOWN UNITS

Reflection

Think of a three-digit number and write it down. Now rearrange its digits to form a new three-digit number. Subtract the smaller from the larger number to find their difference. Finally, circle a nonzero numeral in the difference. Would you believe that I could tell you the number you circled just from knowing the other numbers?

Here's the trick, which you can check for yourself or try on your friends. Take the sum of the remaining numerals in the difference and determine what numeral you would need to add to them to make a multiple of 9; that is the value of the circled numeral. For example, if my original three-digit number were 547, and I rearranged to 745, the difference of these two numbers would be 198. Now suppose I had circled the 8 and told you the other two numerals (1 and 9). You would add those two digits together to get 10 and then determine that my circled numeral is 8 because $10 + 8 = 18$, a multiple of 9. You would also be correct, but why?

The trick works thanks to our base-10 system for representing numbers, as just discussed. However, because the trick is supposed to work for any three-digit number, proving that it works will require us to consider unknown units (discussed further in Chapter 6).

We have considered several different kinds of units and ways that we can operate on those units, such as partitioning and iterating them. Sometimes we want to operate on a unit without even knowing its value. In the present case, we have three unknowns: the three digits in the original number. Let's call them a, b, and c. Each of these digits has an unknown value between 0 and 9. Thanks to our base-10 number system, this number that we might call "abc" is actually $100(a) + 10(b) + (c)$. If we rearrange the numerals, we will get a different number.[10] For example, "bca" would represent $100b + 10c + 1a$. Taking the difference between those numbers, we get $99a - 90b - 9c$.

Notice that 99, 90, and 9 are all multiples of 9. This implies that $99a - 90b - 9c$ is also a multiple of 9 (you might think about why this is so). Therefore, by operating on the three unknown values (a, b, and c) and relying on the base-10 system, we know that no matter what three-digit number you start with, the difference that you end up with will be a multiple of 9. The rest of the trick—determining the value of the digit you circled—depends on a divisibility rule for 9.

To determine whether a number is divisible by 9, we can just add up its digits; if this sum is divisible by 9, the original number also will be divisible by 9. For example, I know that 297 is divisible by 9 and that 332 is not because 2 + 9 + 7 is a multiple of 9 and 3 + 3 + 2 is not. Why does this work? The answer can be found, once again, in our base-10 system.

When we write a number, like 297, in our base-10 system, what we mean is 2(100) + 9(10) + 7(1), but this is the same as 2(99+1) + 9(9+1) + 7(1). More generally, the number "abc" represents 100a + 10b + 1c, which is the same as (99+1) a + (9+1)b + 1c. Distributing a and b, we get 99a + a + 9b + b + c. Notice that 99a and 9b are divisible by 9, so the entire sum will be divisible by 9 if the sum of the remaining terms, a + b + c, is divisible by 9.

By operating on unknown units, we have proved that the trick works regardless the chosen original number. We operated on these unknown units as if they were whole numbers, because they are, only we don't know which whole numbers. As we will investigate further in Chapter 6, this is how algebra works, as generalized arithmetic.[11]

SUMMARY

Generalizing arithmetic to operate on unknown units presents a sizeable challenge for algebra students. This challenge is exacerbated by students' discomfort in operating with other kinds of units, such as composite units and unit fractions. Conversely, if students understand composite units and unit fractions in relation to units of 1, they have a tremendous head start toward algebraic reasoning.[12] Like composite units and unit fractions, unknowns are defined by their relationship to a unit of 1, only with unknowns this relationship is, well, unknown. So, students operating on unknowns must operate on a two-level structure, but students who operate on composite units and students who have been building fractions like 5/3 from unit fractions like 1/3, have been doing that all along.

In general, we find a progression of number sense—from whole numbers to fractions to decimals and generalized arithmetic—that relies on a progression of units and coordinations thereof. As we saw with whole numbers in Chapter 1 and fractions here in this chapter, units themselves arise through the coordination of our own mental actions, such as unitizing, partitioning, and iterating. So, ultimately, the progression of number sense relies on a coordination of mental actions that become more and more complex, with actions on actions. In research on children's development of number, this progression has been framed and tested via the reorganization hypothesis.[13]

The reorganization hypothesis posits that, beginning from counting, children construct numbers from mental actions that they can then reorganize to construct other kinds of numbers. In fact, this is precisely what we have been demonstrating throughout this chapter. In subsequent chapters, in addition to algebra (the subject of Chapter 6), we continue the progression of reorganizations to integers and complex numbers, as directed quantities (the subject of Chapter 8).

Activities

Activity 1: When we refer to a fraction, such as 3/5, we mean 3/5 of a whole—a unit of 1. However, a unit fraction, such as 1/3, can also be a unit. For example, we might want to make 3/5 of a recipe that calls for a 1/3 cup of sugar, so we would need 3/5 of that unit—3/5 of 1/3—and we might want to know what fraction that is of a whole cup. Try to reason through the solution using drawings and the mental actions of partitioning and iterating.

Reflection: Note that your answer is the answer to the multiplication problem, 3/5 × 1/3. Just as taking 3/5 of the whole is 3/5 × 1, taking 3/5 of 1/3 is 3/5x × 1/3. Also, note how the mental actions you used to solve the problem fit those described in this chapter (namely, partitioning and iterating) and how their coordination fits descriptions of fractions multiplication.

Activity 2: Can you think of a situation in which you would ever need to divide two fractions? Consider this one.

Suppose you are making a recipe that calls for 3/4 of a cup of flour, but your only measuring cup is 1/3 of a cup. How many 1/3 cups of flour should you use?

Try to reason your way through it, by partitioning wholes and iterating parts, without relying on the keep–change–flip rule.

Activity 3: Imagine humans had only four fingers on each hand (thumbs included), and that we therefore adopted a base-8 system. Then we might represent numbers using the numerals 0 through 7, with a 1s place, an 8s place, an 8^2 place, and so on (e.g., 23 written in base-8 would represent two 8s and 3 ones, or 19 in base-10). In that system would 7 (being one less than 8) play any special role, like 9 (being one less than 10) does in base-10? Would there be a 7s trick like the 9s trick described in this chapter?

Reflection: There is an additional 9s trick you might have learned, where you raise all ten fingers and put down any one of them (say, the third one). The remaining fingers show the answer to 9 × 3, with two fingers on one side and seven fingers on the other. Why does this work, and how would it work when multiplying by 7 in base-8, with only eight fingers?

Activity 4: Consider the following number trick and explain to yourself why it works:

Think of a whole number between 0 and 9. Now double it four times, and write down the result. Double that number again, and write it down. Now, double one more time and write it down. Finally, add up all the numbers you have written down and subtract your original number. You should now see your original number written three times.

Reflection: By introducing a symbol, such as x, to represent an unknown and then operating on that symbol as if it were a number, you can invent your own trick and see how well it works.

NOTES

1. Note that East Asian countries tend to emphasize part–whole conceptions of fractions less and emphasize measurement conceptions of fractions more. Watanabe (2006) provides a nice contrast between these two curricular approaches, with direct implications for teaching fractions.

2. This phrase comes from Hackenberg (2007). Mathematician and Fields medalist Bill Thurston (1990) recounted the epiphany when fractions become numbers for him:

> I remember as a child, in fifth grade, coming to the amazing (to me) realization that the answer to 134 divided by 29 is 134/29 (and so forth). What a tremendous labor-saving device! To me, '134 divided by 29' meant a certain tedious chore, while 134/29 was an object with no implicit work. I went excitedly to my father to explain my major discovery. He told me that of course this is so, *a*/*b* and *a* divided by *b* are just synonyms. To him it was just a small variation in notation.
>
> (p. 848)

3. The splitting group came about from research on ways that middle school children operate with fractions. Children's splitting operations form the basis for generating this group (Norton & Wilkins, 2012; Wilkins & Norton, 2011).

4. Steffe (2004) referred to equivalent fractions as "commensurate fractions" for this very reason; children understand the equivalence of fractions through measurement concepts of fractions.

5. See Euclid's *Elements* at https://mathcs.clarku.edu/~djoyce/java/elements/elements.html

6. Steffe and Olive (2010) have summarized decades of related research in their book, *Children's Fractional Knowledge*.

7. Note that $P_n P_t = P_{nt}$ represents partitioning and part from a prior partitioning. Steffe (2004) has referred to this composition of partitionings as "recursive partitioning" and has studied its development among children.

8. The Chinese system was base-10, like ours, and relied on physical rods laid out on rectangular arrays, which led to their early invention of linear algebra (Barrow-Green, Gray, & Wilson, 2019), discussed in Chapter 8.

9. Babylonians used base-60 because 60 is a highly divisible number, making it easier for them to work with fractions like 1/4 and 1/15 (Barrow-Green, Gray, & Wilson, 2019. Mayans may have used base-20 because they counted on their fingers and toes.

10.   Note that we are only considering one example of how the digits might be rearranged, but a similar argument works for each of the six possible rearrangements. A more general and rigorous argument might rely on modular arithmetic, which would allow us to focus on the remainders when we divide by 9.

11.   Philipp and Schappelle (1999) describe well the idea of "algebra as generalized arithmetic" in another great resource from *The Mathematics Teacher*.

12.   Research by Hackenberg and Lee (2015) draws strong parallels between operating with fractions and operating on unknowns.

13.   Steffe and Olive (2010) elaborate on this hypothesis in their book.

REFERENCES

Barrow-Green, J., Gray, J., & Wilson, R. (2019). *The history of mathematics: A source-based approach: Volume 1*. Providence, RI: American Mathematical Society.

Hackenberg, A. J. (2007). Units coordination and the construction of improper fractions: A revision of the splitting hypothesis. *The Journal of Mathematical Behavior, 26*, 27–47.

Hackenberg, A. J., & Lee, M. Y. (2015). Relationships between students' fractional knowledge and equation writing. *Journal for Research in Mathematics Education, 46*(2), 196–243.

Norton, A., & Wilkins, J. L. M. (2012). The splitting group. *Journal for Research in Mathematics Education, 43*(5), 557–583.

Philipp, R. A., & Schappelle, B. P. (1999). Algebra as generalized arithmetic: Starting with the known for a change. *The Mathematics Teacher, 92*(4), 310.

Steffe, L. P. (2004). On the construction of learning trajectories of children: The case of commensurate fractions. *Mathematical Thinking and Learning, 6*(2), 129–162.

Steffe, L. P., & Olive, J. (2010). *Children's fractional knowledge*. Boston: Springer.

Thurston, W. P. (1990). Mathematical education. *Notices of the American Mathematical Society, 37*, 844–850.

Watanabe, T. (2006). The teaching and learning of fractions: A Japanese perspective. *Teaching Children Mathematics, 12*(7), 368–374.

Wilkins, J. L. M., & Norton, A. (2011). The splitting loope. *Journal for Research in Mathematics Education, 42*(4), 386–406.

5

Creating, Transforming, and Obliterating Area

n any right triangle with side lengths a, b, and c, where c is the side opposite the right angle (the hypotenuse), $a^2+b^2=c^2$ (see Figure 5.1). This theorem is widely known as the Pythagorean theorem—a misnomer because Pythagoras did not invent the theorem. It was well known to ancient Egyptian, Babylonian, and Chinese mathematicians centuries before Pythagoras was born. Another kind of misnomer appears in the statement of the theorem itself: although it refers to the side lengths of a triangle, the theorem is fundamentally about the areas of squares.

Following the Babylonians, the ancient Greeks referred to 5^2 as 5 *squared* because it represents the area of a square with a side length 5. Likewise, they would call x^3, x *cubed* because it represents the volume of a cube with side length x. So, $a^2 + b^2 = c^2$ should conjure an image of the areas of three squares and the relationship between them. After all, the Pythagorean theorem is essentially about areas; side lengths of triangles are a secondary consideration. Thus, when we consider the Pythagorean theorem, we should primarily concern ourselves with the creation, transformation, and preservation of area.

CREATING AREA

How is area created? We might know how to measure area by determining how many 1-by-1 squares (unit squares) fit into a figure. For example, a 4-by-4 square

DOI: 10.4324/9781003181729-6

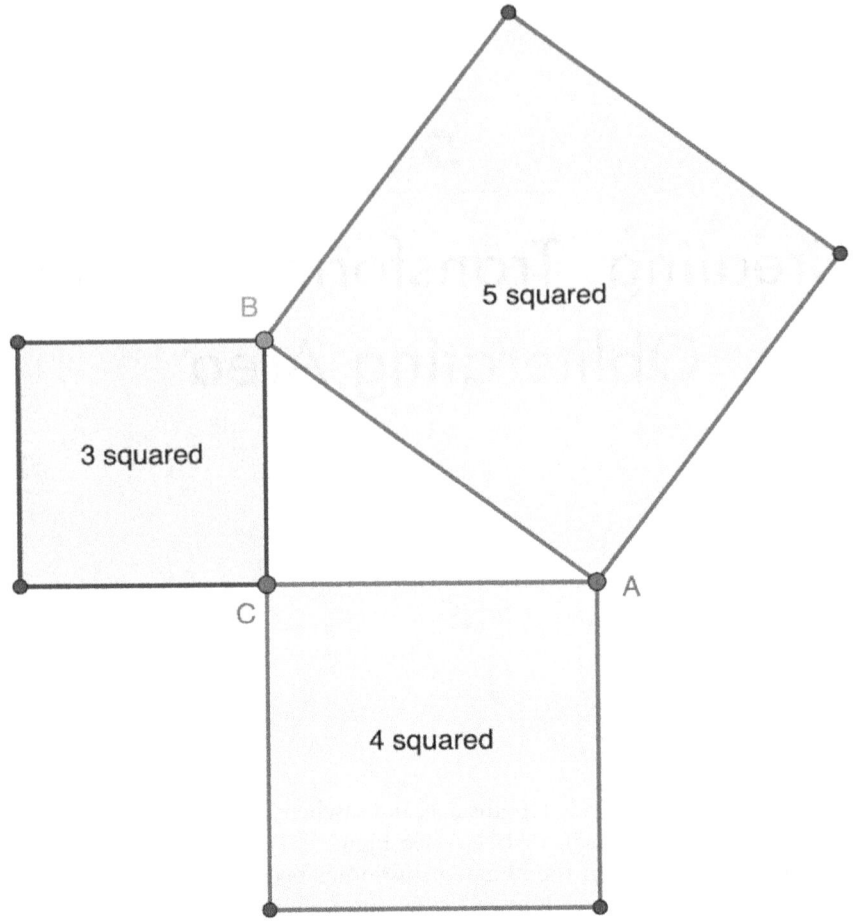

Figure 5.1 Example of a right triangle: $3^2 + 4^2 = 5^2$.

has an area of 16 square units because 16 unit squares fit into it, but how is a unit square created in the first place? Let's start by creating a unit of length.

If you put down your pencil at a point and *sweep* in any direction, you will create a segment of some length. The segment is a trace of the translation of your pencil, in the chosen direction, from one point to another. Its length is a measure of the traced translation, or displacement. We consider length to be one-dimensional because it allows for only one direction of freedom—moving left or right along the segment you drew. Now, you could have chosen to trace out your segment in another direction, but whatever direction you choose produces only a one-dimensional segment.

To increase dimension and produce area, we need to sweep into a second direction, producing a second direction of freedom. Beginning from the line segment,

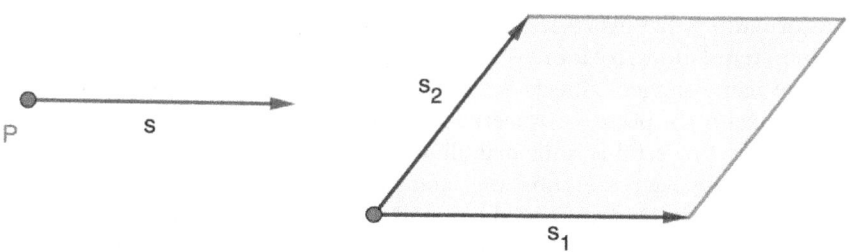

Figure 5.2 Sweeps of points and line segments

which is already the sweep of a point, P, we can sweep out that entire line segment in another direction, s_2, independent of the first, s_1 (see Figure 5.2). Independent means that the second sweep is not in the left or right direction of the first sweep; it isn't in the same line of direction, either forward or reverse. However, we will refer to a degree of independence, indicating the extent to which s_2 moves out of line with s_1.

The area produced by the second sweep depends on its independence from the first sweep.[1] To the degree that it is in line with the first sweep, it will not create area; to the degree it moves out of line with the first sweep, it will. Optimally, the second sweep would move directly out of the line of the first sweep; that is, it would move in a direction that is perpendicular to the first line.[2]

In addition to iterating, reflecting, and partitioning (introduced in Chapters 1, 2, and 4), sweeping constitutes a mental action we can perform to construct mathematical objects. Like Piaget's group of displacements (translations) and Erlangen's principal group, discussed in Chapter 3, there are group-like structures for composing a sweep with other mental actions, including its inverse action, a *projection*.[3]

Whereas a sweep introduces a new dimension, a projection eliminates a dimension. Just imagine a sweep in reverse. For example, we construct area by composing two sweeps, s_1 and s_2 (see Figure 5.2): s_1 sweeps out a line segment from a point, P, and then s_2 sweeps the same length in an independent direction. If we reverse the second sweep, s_2, the area is projected back onto the line segment created by the first sweep, s_1.

TRANSFORMING AND PRESERVING AREA

The mental action of sweeping a line segment produces area, and its inverse action (a projection) annihilates area. So, what mental actions preserve area? We know that these actions include the isometries from Chapter 2 (reflections, translations, and rotations), which preserve both shape and area. What else?

There are many ways we can transform one shape into another shape that has the same area, but fewer ways if we restrict our attention to the kinds of transformations we have considered so far: linear transformations. We treat linear transformations more formally in Chapter 8. For now, we can just think of them as transformations that map lines to lines (or possibly points)—transformations like reflections, translations, rotations, sweeps, projections, and, now, *shears*.

As the name suggests, shears act kind of like scissors, where the vertical and horizontal axes of the plane are squeezed in toward each other (see Figure 5.3). Shears are cognitively powerful actions that allow us to understand why many different geometric figures have the same area, and as we will soon see, shears undergird Euclid's proof of the Pythagorean theorem. Like other geometric transformations, shears transform the entire plane, not just geometric figures. However, it's helpful to use a square as a frame of reference to understand how shears transform the plane. Consider the example shown in Figure 5.4.

Reflection

Why doesn't the shear shown in Figure 5.4 change area?

Looking from the left side to the right side of Figure 5.4, the left side of the square stays the same, but the right side of the square is shifted vertically while remaining parallel to the left side of the square. The area swept out by the bottom side also remains the same; it's swept up by the length of the left and right sides, which haven't changed. The length of the bottom side has changed but only by introducing a component that is in the same direction as the sweep, which introduces no new area. We can also make an argument based on the figures by noting that the right triangle lost from the bottom of the square is gained at the top of the parallelogram, but keep in mind, figures are of secondary importance. Throughout this chapter (and the book as a whole), we try to stay focused on the transformations themselves.

Figure 5.3 Shearing the plane.

Source: © Eleanor Norton

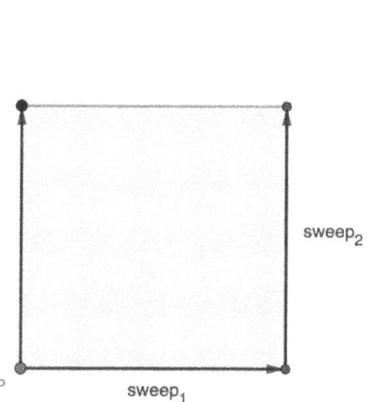
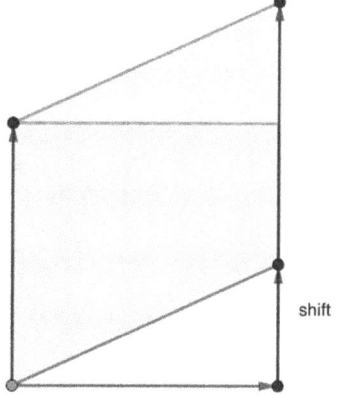

Figure 5.4 Shearing a square.

A TRANSFORMATIONAL PROOF OF THE PYTHAGOREAN THEOREM

The following proof is grounded in two of the mental actions that we have been discussing: sweeps and shears. We will see how the Pythagorean theorem arises from a coordination of those mental actions, in creating, transforming, and pre-serving area. After demonstrating the proof, we will see evidence that the same kinds of transformations undergird Euclid's original proof of the theorem. Then, we will consider how the coordination of those transformations, as mental actions, might lead to an understanding of the Pythagorean theorem in one fell swoop.

We begin our proof by drawing a square of area a^2—the result of sweeping a line segment of length a perpendicularly by a distance of a. Then, to create the conditions under which the theorem applies (perpendicular segments of length a and b), we shear the square as shown in Figure 5.5. This creates the right triangle with side lengths a, b, and a third side, labeled with length c. In shearing the square, we preserve the area, a^2.

Recall, we are considering the Pythagorean theorem as a statement relating three squares: a^2, b^2, and c^2. We have already generated an area of a^2. The next few transformations will help us compare the areas a^2 and c^2. First, we transform the par-allelogram with area a^2 into a rectangle by shearing the parallelogram as indicated in Figure 5.6. Next, we complete c^2 by sweeping the topmost segment of the rect-angle by some unknown additional length, producing a square with side lengths c and area c^2 (Figure 5.7).

At this point, a^2 is represented by the area of the rectangular area shown on the left side of Figure 5.7, and c^2 is represented by a larger area shown on the right side of Figure 5.7. The new area represents the difference between c^2 and a^2 (the

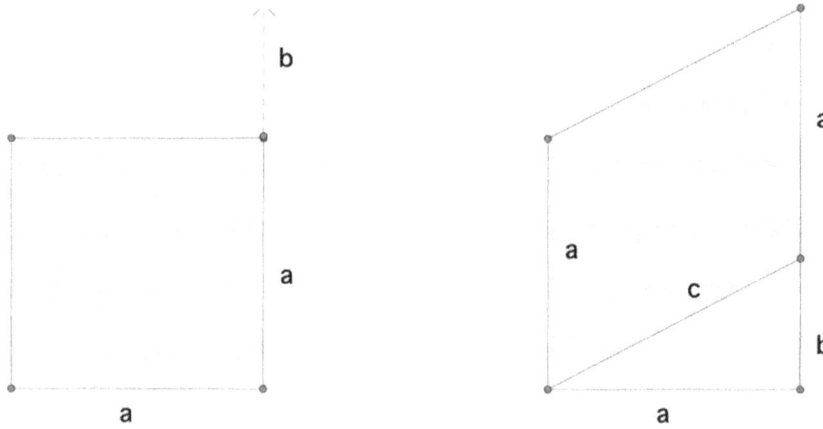

Figure 5.5 Sweeping and shearing area.

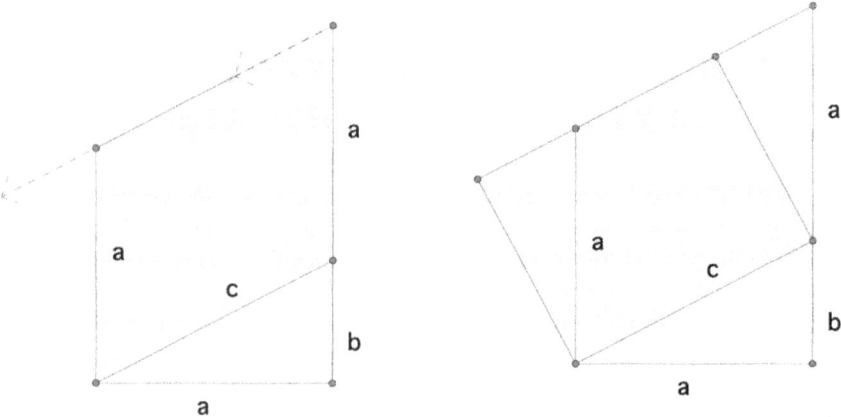

Figure 5.6 Shearing the parallelogram into a rectangle.

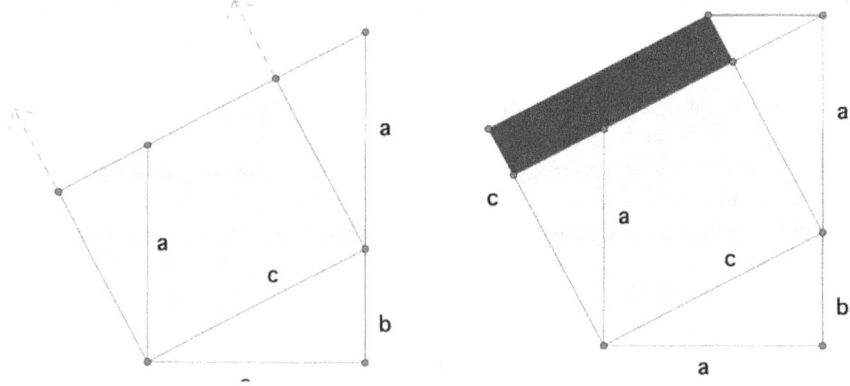

Figure 5.7 An additional sweep

additional area that is swept). To prove the Pythagorean theorem, we need to prove that that difference is b^2.

Notice that the same line segment, of length a, appears at the bottom and right sides of the figure. At the bottom, c is formed by shifting the right endpoint of that line segment perpendicularly by a distance of b. The same is true for the length, a, appearing on the right side of the figure: the length c is achieved by shifting top endpoint perpendicularly by distance b (see the left side of Figure 5.8).

Reflection

Try to rehearse the mental actions that transform the square with side length, a, and area, a^2 (from Figure 5.4), into the rectangle shown on the left side of Figure 5.7. Similarly, try to rehearse mental actions demonstrating that the larger rectangle, on the right side of Figure 5.7, must have area c^2. What about the area of the new rectangle?

Finally, we can perform a pair of shears to produce a square with area b^2 from the new rectangle. We first shear the rectangle along the line segment indicated in Figure 5.8. Because this shear is the reverse of the shear shown in Figure 5.6, we know that it produces a parallelogram having a pair of opposite sides with length b. Specifically, we know that the top side of the parallelogram will coincide with the segment highlighted on the right side of Figure 5.7. The second shear transforms the parallelogram into a square, as illustrated in Figure 5.9. We know the shear results in a b-by-b square because its width is b, and the total height of the figure is a + b, leaving a height of b for the square.

This transformational proof relies on a sequence of mental actions—sweeps and shears—to create, transform, and preserve area. Through these actions, we find that the original area, a^2, can be created through a pair of sweeps and preserved through a pair of shears and that the additional area needed to complete c^2 is b^2, as

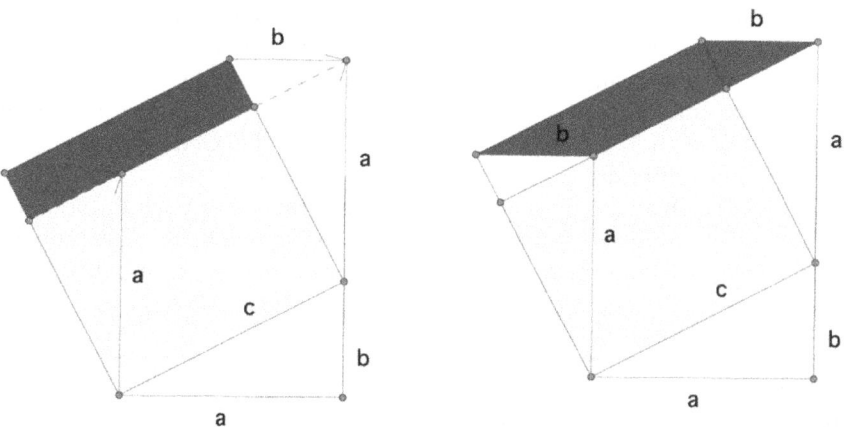

Figure 5.8 Shearing the new rectangle into a parallelogram.

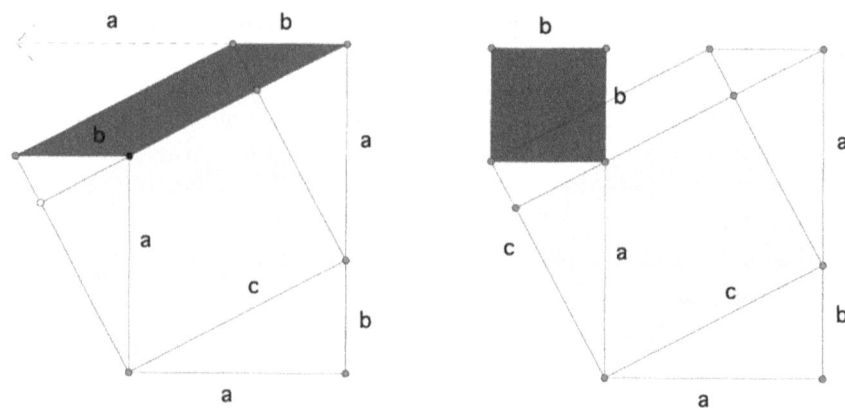

Figure 5.9 Shearing the parallelogram into a square.

demonstrated by another sweep and two more shears. Although Euclid's original proof of the theorem is much more formal, we can see evidence of these same mental actions undergirding his long chain of arguments.

EUCLID'S PROOF

Euclid was the first mathematician to provide a formal foundation for mathematics. In his book of *Elements*, he laid out an axiomatic system for geometry—a system that is useful in communicating logically rule-based arguments but one that belies the dynamic nature of mathematics and one's individual power to construct it. Still, we will see evidence that Euclid himself reasoned transformationally to build his formal arguments.

The purpose of the *Elements*, Book I, was to prove the Pythagorean theorem from common notions and postulates (axioms). The first three axioms (Postulates 1–3; see Table 5.1) are Plato's rules for straight edge and compass constructions, indicating their sensorimotor basis within Greek culture. They describe how you can draw/define lines and circles from pairs of points, and how you can make new points where lines and circles intersect. The chain of propositions leading from those axioms to the Pythagorean theorem indicates the kinds of mental actions behind Euclid's intuitions.

The Pythagorean theorem appears as Proposition 47 in Euclid's *Elements*, the penultimate proposition of Book I, followed by its converse. The proposition immediately preceding the Pythagorean theorem concerns constructing a square

Table 5.1 Euclid's first three postulates[4]

Postulate 1	To draw a straight line from any point to any point.
Postulate 2	To produce a finite straight line continuously in a straight line.
Postulate 3	To describe a circle with any center and radius.

area from a given line segment (Proposition 46). Euclid's proof of the Pythagorean theorem relies heavily on that proposition and propositions about shearing.

The left side of Figure 5.10 illustrates the diagram Euclid used to support his arguments for the Pythagorean theorem. Essentially, he argued that the areas of the two rectangles making up the large square, with area c^2, were equivalent to the areas of the two small squares, with areas a^2 and b^2. The argument depended on previous propositions that appealed to shearing (Propositions 35–38, & 41). In the figure, triangle DAC has half the area of the parallelogram that results from shearing rectangle DAGF along segment FC. Because triangle DAG has the area of half of that rectangle, triangles DAC and DAG have the same area. Likewise, triangle BAE has the same area as CAE. Because triangles DAC and BAE are congruent (by side–angle–side congruence), the areas of the smaller rectangle and smaller square (each having twice the area) are equivalent. The same argument works for the larger rectangle and the larger square, thus proving the Pythagorean theorem.

The Pythagorean theorem was the focus and main prize of Book I, with the preceding postulates building up to it. Euclid interpreted the theorem as we did, as a relationship between the areas of three squares rather than an equation relating the side lengths of a triangle.[5] He decomposed the largest square like we did and then transformed the decomposed figures while preserving their area. The power of Euclid's approach is evident in the sheer number of propositions that he proved along the way. The power of the transformational approach lies in an appeal to intuition and our own personal power in constructing mathematics. The connection between the two approaches elucidates the dynamic underpinnings of formal mathematics.

A DYNAMIC DEMONSTRATION

The right side of Figure 5.11 displays our transformational proof, with vertices A, B, and C relabeled to match Euclid's labels. Note that the same pair of rectangles appear in each diagram. This comparison underscores the argument that Euclid's

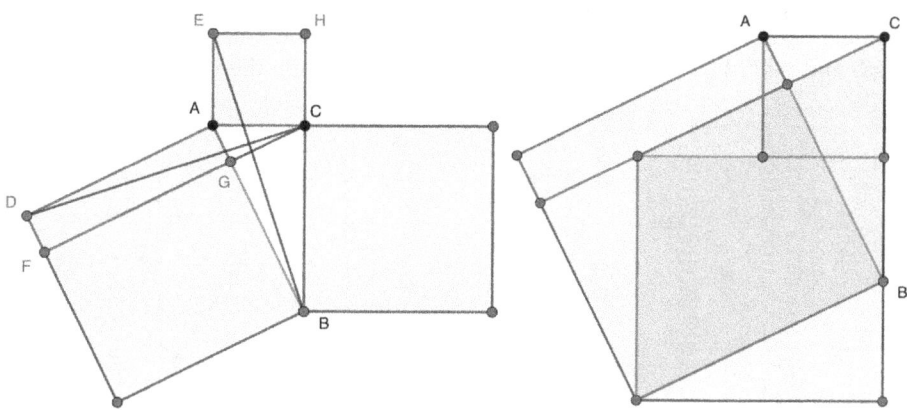

Figure 5.10 Comparing proofs of the Pythagorean theorem.

proof was motivated by a coordination of mental actions similar to the ones presented in our transformational proof.[6] Now, we consider how that coordination generates the Pythagorean theorem all at once.

Euclid relied on geometric constructions of static figures to produce a long chain of propositions (46 to be exact) leading to the Pythagorean theorem. Each proposition built on previously demonstrated propositions, ultimately leading back to Plato's three rules of geometric construction (see Table 5.1).[7] The three rules pertain to sensorimotor actions of drawing lines and swinging circles, and we have argued that the propositions too, rely on actions like sweeping and shearing. It's the coordination of these dynamic entities—not the static figures—that concern us now.

From this perspective, the Pythagorean theorem simply reminds us that area is created through perpendicular sweeps. The square, a^2, is produced by sweeping a line segment of length a by another length a, in the perpendicular direction. When each length a is transformed into a length of c by introducing a perpendicular component, b, the c-by-c sweep introduces the additional area of b^2. We get an additional area of b^2 because that is exactly the area created by the perpendicular sweeps of length b that transform length a into length c (see Figure 5.11).

The left side of Figure 5.11 shows the original a-by-a square with area a^2. As we introduce perpendicular components of length b, to the bottom and right sides (see the center and right sides of Figure 5.11), we get a new square with area c^2. In the process, we are sweeping out an additional area of b^2 (the small square) because one side of length c has a component of length b that is perpendicular to another component of length b, within the other side of length c.

Reflection

Rehearse the mental actions described above until you can see the square area being created as point A is shifted away from point C.

We can generalize this argument to rectangles (see Figure 5.12). Starting with a rectangle with width a and length ka (where k is the ratio between the rectangle's length and width), we can shear it vertically by b. Righting the rectangle requires a

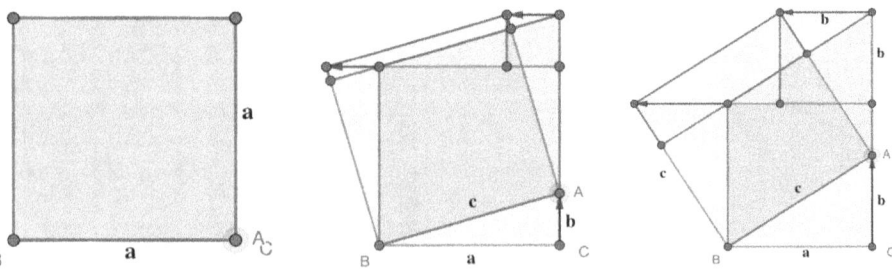

Figure 5.11 A Pythagorean transformation.

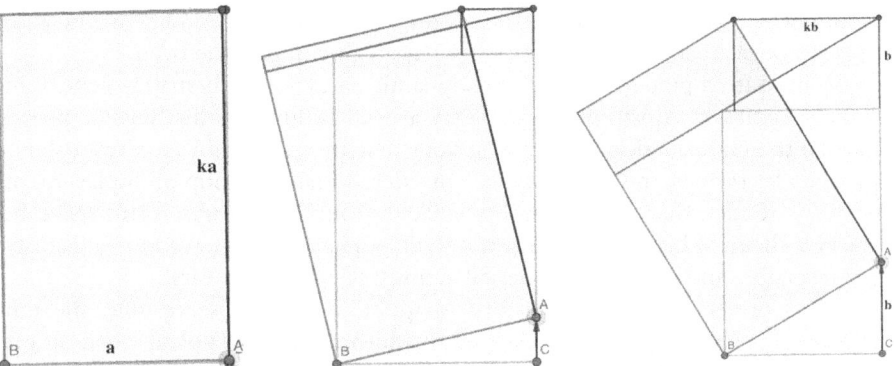

Figure 5.12 A generalization of the Pythagorean theorem.

horizontal shift of kb, and the additional area swept out will be b × kb, as represented by the new rectangular areas at the top of Figure 5.12.

We can make the argument using the same shears that undergird Euclid's formal proof of the Pythagorean theorem, or we can appeal to the single sweeping operation that arises from their coordination. Namely, the additional area is created by the perpendicular sweeps, vertically by b and horizontally by kb. As a mathematical operation, this mental action is also reversible, meaning that if the large rectangle were "unrighted" by a shift of kb in the opposite direction, the amount of area lost would be the same. The power of this reversible operation will become more evident in Chapter 8 in demonstrating how determinants of matrices determine the way they transform of area (through matrix multiplication).

The purpose of the present chapter was to demonstrate that formal mathematics—even Euclid's axiomatic proof of the Pythagorean theorem—is accessible to us through the coordination of our own mental actions. Indeed, we have argued that Euclid's own mental actions of sweeping and shearing undergirded his formal proof. We can learn to coordinate such actions to reach an understanding of mathematical ideas, like the Pythagorean theorem, in one fell swoop. If we work at it, eventually we can just see it, like the seven yellow bricks in Chapter 1.

SUMMARY

You have probably seen the Pythagorean theorem many times before and might have seen a formal proof of the theorem, presented as a logical sequence of arguments, beginning from axioms, like the proof and axioms that appear in Euclid's *Elements*. This is the way we usually see mathematics, in textbooks and in classrooms, in its final crystallized form.[8] It's a bit like attending a baking class in which you are presented with cupcakes that you might consume (if you like that sort of thing) but without ever actually baking. Yes, cupcakes are the end goal, but you'll never become a baker if you never see what's happening in the oven.[9]

We have attempted to elucidate the dynamic nature of mathematics—the doing of mathematics. In constructing geometry, as well as number, the dynamic process relies on the coordination of mental actions that are composable and reversible, as described by groups. In Chapter 3, we considered Piaget's group of displacements, and we extended that group to the principal group, which transforms space but preserves shape. While investigating the Pythagorean theorem, we introduced two new reversible and composable mental actions: sweeps and shears.

Shears preserve area but transform shape. They also preserve lines, meaning that every line in the plane gets mapped to another line in the plane, without getting bent or broken. In Chapter 8, we consider the group of such transformations: linear transformations of space that preserve area. We can represent this group as 2-by-2 matrices whose determinants are 1 or -1. We also consider matrices that transform area and how determinants describe this transformation.

Sweeps are transformations that can create area. They are traces of displacements, so area is created by composing components from the group of displacements. In our investigations of the Pythagorean theorem, we found that the two traced displacements (sweeps) create area only to the degree that they are independent. When perpendicular components of length b were introduced to transform sides of length a into sides of length c—thus transforming a^2 to c^2—an additional area of b^2 was produced, that is, $a^2 + b^2 = c^2$.

Sweeps are reversed by parallel projections. Whereas sweeps create new dimensions (creating a length by sweeping a point, creating area by sweeping a length in a new direction, or creating volume by sweeping an area in yet another direction), projections destroy dimensions. We might consider the group formed by these two kinds of action, but their various compositions would produce unlimited dimensions. Note that such a group would be quite different from Klein's group of projective transformations (as mentioned at the end of Chapter 3).

Activities

Activity 1: Beginning with a line segment, can you use Plato's three rules (also known as Euclid's first three postulates; see Table 5.1) to construct a perpendicular line? Can you then construct a square?

Activity 2: How might we generalize the Pythagorean theorem to three dimensions? In particular, if you have a box with length l, width w, and height h, how far will its opposite corners be?

Activity 3: Wasserman, Weber, Fukawa-Connelly, and Mejía-Ramos (2020) suggest a mental action akin to shearing that would make it easy to compute the areas of various geometric figures.[10] Begin by choosing one dimension, such as the vertical diagonal of the kite shown on the left side of Figure 5.13.

Then shift the entire figure perpendicular to that dimension. Just imagine the point P sliding along the horizontal diagonal to point Q so that the vertical diagonal is all the way at the end, forming a triangle. As with shearing, this shift should not alter the total area of the figure but can make it easier to find that area. Use this action to find the area of the kite. Use the same action to explain why the area shown on the right side of Figure 5.13 is 12.

Activity 4: The area of a large parallelogram, shown in Figure 5.14, is the area of the large rectangle less the area of the small rectangle. Why is that? The generalization of the Pythagorean theorem, illustrated in Figure 5.12, might help you. We'll return to this idea in Chapter 8 while investigating the determinants of matrices.

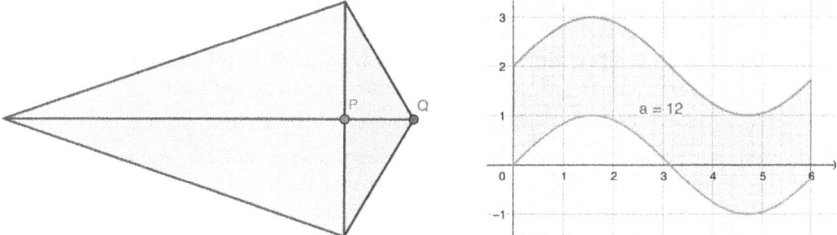

Figure 5.13 Area-preserving transformations

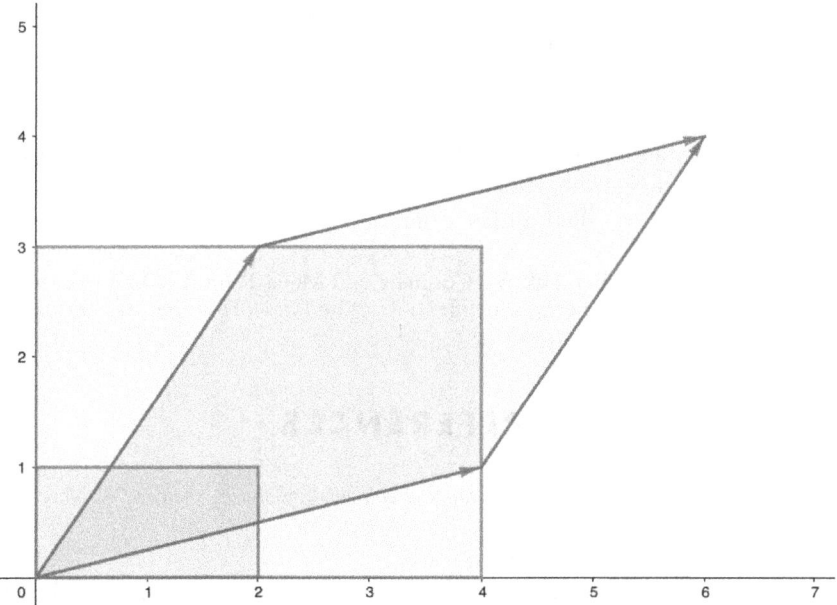

Figure 5.14 Area of a parallelogram

NOTES

1.   See Thompson (2000) for an exposition on area as "independent directions of sweep."

2.   Children have an intuitive sense of this direction, but it is one they must learn to abstract. When I asked my daughter Caroline, at 5 years old, to walk directly toward a flat wall from the middle of a room, she walked in a direction perpendicular to the wall. She described the path as one that went "straight to" the wall, and she could connect this direction to the one that gave the shortest distance to the wall. My older daughter, at 8 years old, could do more: she could also find the closest point on a drawn line from a drawn point. Studies of children's conceptions of angle measure have found a related developmental stage as 6- and 7-year-old children perceive angle measures but struggle to represent them in drawings (Piaget & Inhelder, 1967).

3.   Here we refer to a parallel projection, and not kind of projection used in projective geometry (emanating from a point), mentioned at the end of Chapter 3. We also distinguish it from the kind of projection mentioned in the Introduction, where we project our structures into the universe.

4.   You can find all of Euclid's postulates organized within this excellent resource: https://mathcs.clarku.edu/~djoyce/java/elements/bookI/bookI.html

5.   See Maor (2007) for more on the history of the Pythagorean theorem.

6.   We have noted Euclid's reliance on sweeps and shears. He seemed to take a third mental action, translation, for granted in arguing that figures that coincide are congruent: "Euclid failed to consider displacements as geometric operations, although he used them continuously" (Beth & Piaget, 1966, p. 293).

7.   "To deduce, is to construct. Only hypothetical judgments are demonstrable; we demonstrate that one thing is the consequence of another. As a result we construct the consequence with the hypothesis. The conclusion is necessary, although it involves novelty, not because it is contained in the hypothesis, but because it is derived by fixed operations, that is to say, which are not arbitrary. And what rules do these operations follow? Those of formal logic? Certainly not, they are rather propositions already accepted, either in virtue of preceding demonstrations, or as definitions and postulates. The application of these propositions to the constructive operations is precisely the function and role of the syllogism in reasoning." (Goblot, 1922, pp. 50–51; as cited in Beth & Piaget, 1966, p. 19).

8.   See Lakatos's (1976), *Proofs and Refutations.*

9.   Paul Lockhart (2009) makes a similar analogy, between math and music, in *A Mathematician's Lament.*

10.   Wasserman, Weber, Fukawa-Connelly, and Mejía-Ramos (2020) refer to a two-dimension version of Cavalieri's principle to describe "area-preserving transformations."

REFERENCES

Beth, E. W., & Piaget, J. (1966). *Mathematical epistemology and psychology* (W. Mays, Trans.). New York: Gordon and Breach.

Goblot, E. (1922). *Le système des sciences.* Paris.

Lakatos, I. (1976). *Proofs and refutations: The logic of mathematical discovery.* New York: Cambridge University Press.

Lockhart, P. (2009). *A mathematician's lament*. New York: Belleview Literary Press.

Maor, E. (2007). *The Pythagorean theorem: A 4000-year history*. Princeton, NJ: Princeton University Press.

Piaget, J., & Inhelder, B. (1967). *The child's conception of space* (F. J. Langdon & J. L. Lunzer, Trans.). New York: Norton (Original work published in 1948).

Thompson, P. W. (2000). What is required to understand fractal dimension? *The Mathematics Educator, 10*(2), 33–35.

Wasserman, N. H., Weber, K., Fukawa-Connelly, T., & Mejía-Ramos, J. P. (2020). Area-preserving transformations: Cavalieri in 2D. *Mathematics Teacher: Learning and Teaching PK-12, 113*(1), 53–60.

6

The Power of Symbols

In Chapter 1, we considered how children construct numbers and how those constructions explain immutable facts, like $2 + 2 = 4$. As long as we agree on the actions behind the symbols "2" "+" and "4", the equality amounts to nothing more than tautology—two different ways of representing the same composition of units. Specifically, we can understand addition as the continuation of a count (the iteration of a unit of 1) wherein 2 and 4 have their respective places. Both $2 + 2$ and 4 represent $1 + 1 + 1 + 1$, counted as $1, 2, 3, 4$.

As von Glasersfeld explained, numerals like 2 and 4 call to mind our acts of counting.[1] They serve as symbols that re-present our actions and allow us to act on them further. Because of our familiarity with them, we often overlook numerals as symbols. They are among the first mathematical symbols we use because numbers are among the first mathematical objects we construct, and constructing mathematical objects is a necessary precursor for meaningfully symbolizing them.

Of course, we use symbols all the time, in all walks of life. Each word you read now is a symbol that stands in place of some referent or some relationship between referents. However, mathematical symbols not only communicate ideas; they serve as proxies for mathematical objects. As such, we can act on these symbols as if they were the mathematical objects they represent, just as we did with the geometric figures in Chapter 5. We do something similar in chemistry and cooking when we combine elements or ingredients. We might even balance chemical equations, as

DOI: 10.4324/9781003181729-7

in $2H_2 + O_2 = 2H_2O$. However, these equations work out only to the degree that mathematics applies them.

In mathematics, the objects themselves are coordinations of actions. Thus, we can combine mathematical symbols as a proxy for composing mental actions, producing new mathematical objects, as in producing 4 from 2+2. In that way, we can make sense of numbers we have never even considered before. For example, we can appreciate 257×351 as a number even before computing this product and naming it.

Prior chapters illustrate how your mind makes mathematics. That is, they demonstrate how you might construct mathematical objects, such as numbers and shapes, from coordinations of your own mental actions. Constructing mathematical objects comes with advantages. These objects become available for you to project into the world by structuring and organizing your experiences within it. Now we consider the additional power that comes with symbolizing those objects.

COMPUTING

Mental arithmetic provides an ever-ready opportunity for us to exercise our own mental actions. We might ask our children to compute sums and products in their heads, and we often become frustrated when they reach for a calculator (or phone) instead. We have an inherent sense that some mental math would do them good. We might even expect them to work out products like 31×29 and encourage them to use strategies like thinking about the product as $(30 + 1) \times (30 - 1)$, which is $900 - 1$. However, regardless of such strategies, at some point, computations become overwhelming, and at that point, we rely on symbols and algorithms as proxies for coordinating the underlying mental actions that they represent.

In Chapter 4, we considered the mental actions of partitioning and iterating, which students might coordinate in order to construct fractions. In the case of multiplying and dividing fractions, this coordination can become very complicated. For example, consider the following task:

> You are at a party and a cake is cut into nine pieces. Two people show up to the party late and you decide to share your piece of cake with them. What fraction of the whole cake do the latecomers get together?[2]

You might understand this task as a fraction multiplication task and, thus, convert it into the exercise of computing $2/3 \times 1/9$. Then you could apply the multiplication algorithm (multiplying the numerators and the denominators) to obtain $2/27$. There is great economy in approaching the problem that way if you understand what the symbols and the algorithm represent. Understanding what they represent means, first, understanding fractions as numbers made from your own coordinated activity of partitioning whole units and iterating fractional units. Then, it means operating on those numbers, as mathematical objects, again using mental actions like partitioning and iterating.

Consider the diagram shown in Figure 6.1, which illustrates the mental actions you might perform to solve the cake task. You might begin by producing a mental image of one-ninth of the whole cake, by partitioning it into nine equal parts (P_9) and taking out (disembedding) one of them (D_1). Now, to find one-third of that piece, you might partition it into three equal parts (P_3) as if it were a new whole. Disembedding one of those parts (D_1), you have a piece that is "1/3 of 1/9" of the original whole. To determine how big that piece is in comparison to the whole, you might distribute three parts into each of the nine parts that make up the original whole ($T_{3:9}$). Thus, you can see that 27 of that new piece fit into the whole, rendering the piece 1/27 of the whole. Finally, you need two of those pieces, to make *two* thirds of 1/9, so you might iterate the piece twice (I_2), producing 2/27 of the original whole.

Reflection

Looking at the diagram in Figure 6.1and reflecting on your own mental actions, can you justify why we multiply numerators and denominators when multiplying two fractions?

For symbolic manipulation to have meaning, the symbols and their operations (e.g., fractions multiplication) must ultimately refer to our own mental actions. Otherwise, we are reduced to meaningless symbol manipulation found all too commonly in our classrooms.[3] On the other hand, if we had to go through the mental actions every time we computed something, without relying on algorithms, we would quickly become overwhelmed. Therefore, the goal is for us to make meaning of symbols and algorithms on the basis of our own mental actions so that symbolic manipulation can stand in place of them and help us keep track

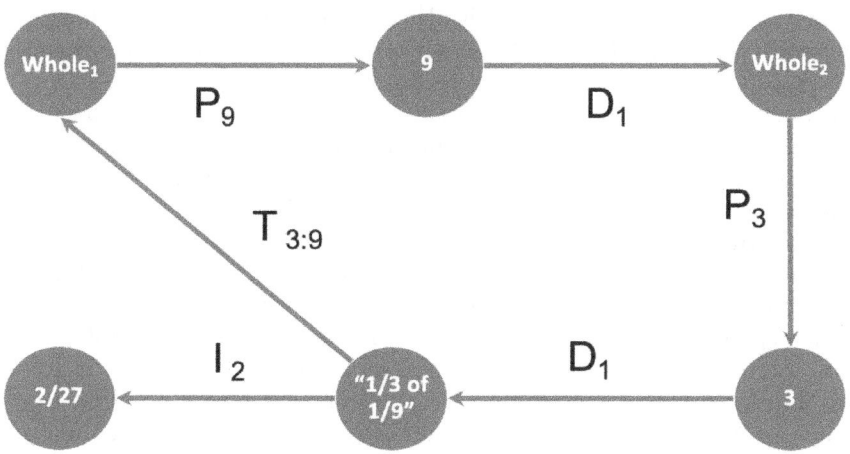

Figure 6.1 Mental actions for multiplying fractions.[4]

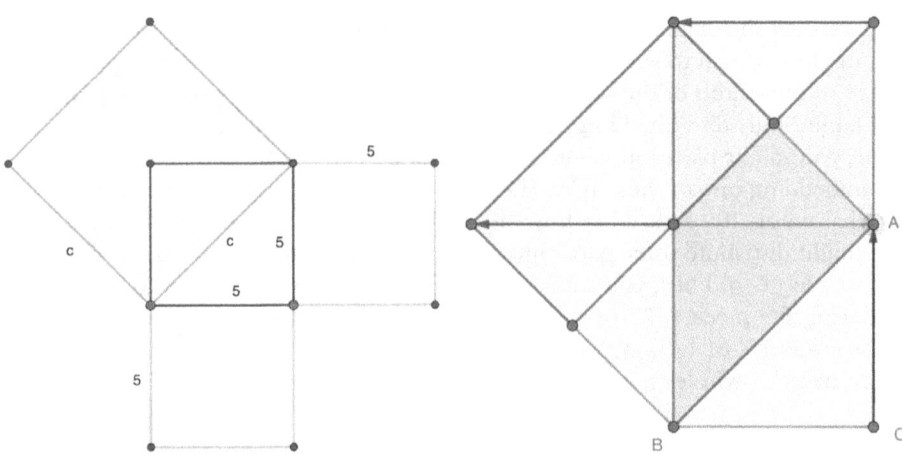

Figure 6.2 Applying the Pythagorean theorem

of them. For example, the keep–change–flip rule (discussed in Chapter 4) can be a time-saving procedure for performing fraction division, but when it is taught as a meaningless rule, students might as well use a calculator app, available on virtually every mobile device.

DESCRIBING GEOMETRIC RELATIONSHIPS

When symbols stand in place of geometric objects, you can manipulate those symbols as a proxy for geometric constructions. For example, you can represent a square with a side length 5 as 5^2. After all, 5^2 symbolizes the product of two sweeps that produce the square (sweeping a line segment of length 5 and then sweeping that line segment by 5 in a perpendicular direction), as described in Chapter 5. Now you might want to find the length of a diagonal in that square, and the Pythagorean theorem tells you how (see Figure 6.2).

Bringing the Pythagorean theorem to bear on a square and its diagonal, as shown on the left side of Figure 6.2, allows us to formulate the image on the right without having to recreate it. The image on the right represents a problem we have already solved in generality (in Chapter 5), resulting in the equation $a^2 + b^2 = c^2$. In the present case, we have $5^2 + 5^2 = c^2$, where c is the diagonal of the original square, which we can determine with a little calculation and manipulation.

In other words, we have already gone through the labor of constructing squares from sweeps and deriving the Pythagorean theorem through the coordination of additional mental actions, such as shearing. Once it is symbolized, there is no need to go through the process every time. We can offload the work onto the symbols through symbolic manipulation. This is just a simple application of the most powerful tool in all of mathematics: algebra.[5]

OPERATING ON UNKNOWNS

Sometimes referred to as generalized arithmetic, algebra involves working with unknown quantities as if they were numbers.[6] Operating on unknown quantities requires us to hold in mind a relationship between that unknown quantity and a unit of 1, similar to operating on a composite unit—operating on two levels at once.[7] We can represent this unknown relationship with a symbol, like x. We can then rely on algebraic manipulation of this symbol with other symbols (including numerals) to add, subtract, multiply, or divide. These operations stand in place of mental actions we might perform on the unknown quantity, such as iterating and partitioning. For example, consider the following problem from ancient Egypt:

A quantity and its 1/7 become 19. What is the quantity?[8]

Egyptian mathematicians solved problems like this thousands of years before algebra was invented, and they had to be especially clever to do so. In this case, they were challenged to posit an unknown number, to add it to a 1/7 part of that same unknown number, and to equate it with 19. In other words, they had to coordinate the mental actions of unitizing a quantity, partitioning that unit into seven equal parts, and combining the unit with one of those parts, knowing that the result is a composite unit of 19 ones but without knowing the value of the original quantity in units of 1.

Reflection

How would you solve this problem if you could not use algebra?

The Egyptian method was called "false position," which was a sophisticated form of guess-and-check. Using this method, we might guess that the original quantity is 14 because it's pretty close to 19 and because it's conveniently divisible by 7. This guess would yield $14 + 2 = 16$, which is not far off from 19. We would then adjust this initial guess. Noting that 16 is 3 away from the desired sum, we might add amounts to the first guess of 14—amounts that would add 2 and 1 to the sum (16). Specifically, 2 is half of half of half of 16, so adding half of half of half of 14 $(1 + 1/2 + 1/4)$ to 14 will yield $16 + 2 = 18$ in the sum.[9] Adding another half of $1 + 1/2 + 1/4$ $(1/2 + 1/4 + 1/8)$ will add 1 more to the sum, yielding the desired result, 19. So, the answer is $14 + 1 + 1/2 + 1/4 + 1/2 + 1/4 + 1/8 = 16 + 1/2 + 1/8$.

Using algebra, the problem becomes much more straightforward. Once we symbolize an unknown quantity, we can use algebraic manipulation as a proxy for acting on that quantity. In the present case, we might symbolize the unknown quantity with an *x*. We don't know the value of x, but we can act on it anyway, symbolizing those actions too. Specifically, we want to consider the unknown quantity, x, and its one-seventh, which would be x/7, or 1/7 x, where "1/7" symbolizes the mental action of partitioning the unknown quantity into seven equal parts and taking one of them. Together, this makes x + 1/7x, where "+" symbolizes the action of combining (unitizing) the two quantities to form a new quantity. Now we can equate this new quantity with 19, using the equation x + 1/7x = 19.

We have symbolized the problem by symbolizing various actions (e.g., partitioning and unitizing) on an unknown quantity. Now, to determine the value of that quantity, x, we have to undo those actions, and this is possible because mathematical mental actions are reversible. We can separate x + 1/7x into its two parts, x and 1/7 x, and consider what happens when we iterate each part 7 times (undoing the mental action of partitioning x). The result is 7x and x. So, together, 7x and x should be 7 iterations of 19, which is 7 × 19 = 133. We have thus simplified the problem to 7x + x = 133, or 8x = 133. If 8x is the quantity 133, then partitioning that quantity into eight equal parts should give the value of x. Indeed, x = 133/8, or 16 5/8.

Consider how difficult it would be to keep in mind all the mental actions that built up the equation and all the actions we took to reverse those actions and determine the value of the unknown quantity. This is the power of algebra. It allows us to offload the cognitive demands required to coordinate all those mental actions. We use the algebraic symbols and their manipulation as a proxy for all that work. This power becomes ever more profound as we work with ever more complicated and extended coordinations. In Chapter 4, we saw another example demonstrating the power of representing and operating on unknown quantities, with the 9s trick and our base-10 system.

Unknowns hold particular but unknown values. In contrast, true variables vary, like the temperature. When they vary, other variables might vary with them. Consider the equation y = 9/5x + 32. This equation tells us nothing about the particular values of x or y, only about a relationship between them. It tells us how they covary. Specifically, it represents a linear relationship between temperature measured in degrees Celsius and temperature measured in degrees Fahrenheit.

One thing we can tell right away is that when x gets big, *y* gets even bigger. The x value is transformed first by partitioning it into five equal parts, then by iterating that part 9 times, and finally by combining it with a value of 32. While the first transformation will make positive values smaller, the second transformation more than makes up for that, by iterating the one-fifth part beyond the five-fifths whole, and the third transformation contributes additional value to the result. Together, these transformations tell me how x is transformed into y, and therefore, how x and y covary.

Because x is transformed into *y*, we might represent the equation, y = 9/5x + 32, as a function, f(x) = 9/5x + 32. In this notation, *f* represents the function, and *x* represents the independent variable, or the variable that we consider to vary freely within some specified domain (e.g., x might take on the value of any real number). It's tempting to think about the function as a kind of machine that takes in arbitrary values of x and spits out associated values of y, but this metaphor belies the nature of variables. To fully appreciate functions, especially continuous functions like f(x), our minds must allow the two variables to covary, the subject of Chapter 9.

DERIVING FORMULAS

To see the full power of representing and operating on unknown quantities, let's consider a more complicated equation: the quadratic formula. As mentioned in Chapter 5, when the ancient Babylonians and Greeks referred to a squared quantity (e.g., 5^2), they literally referred to a square with the given side length (5). Therefore, they used geometry to solve problems that, to us, might look like quadratic equations (equations of the form $ax^2 + bx + c = 0$, where a, b, and c are integers or fractions). For example, consider the following problem from an ancient Babylonian tablet:

I have added the area and two-thirds of the side of my square and it is 35/60. What is the side of my square?[10]

Using algebra, we could represent this problem as $x^2 + 2/3x = 35/60$. However, living in 2000 BCE, these ancient Babylonian mathmaticians did not have algebra, which wasn't fully developed until the 16th century CE. So, they would solve the problem as follows:

You take 1, the coefficient. Two-thirds of 1 is 40/60. Half of this, 20/60, you multiply by 20/60 and the result 6/60 + 40/3600 you add to 35/60 and the result 41/60 + 40/602 has 50/60 as its square root. The 20/60, which you have multiplied by itself, you subtract from 50/60, and 30/60 is the side of the square.

Figure 6.3 illustrates the solution. As described in the problem, we have a square and 2/3 of its side length, which we can represent with a rectangle with a width of 40/60 and a length that is the same as the square's unknown side length. Now we take "half of this" 2/3, in order to split the rectangle into two equal rectangles. We align one of those rectangles on the right side of the square and the other rectangle, rotated, with the bottom of the square. This forms a new square but with one corner missing, so we are going to *complete the square*. That is precisely what the tablet describes in saying "20/60 [or 1/3], you multiply by 20/60 [i.e., multiply 1/3 by itself or *square* it]." If we add the area of this new square (6/60 + 40/3600; known to us as 1/9) to the original area (35/60, which is the area of the original square added to the area of the original rectangle), we now have the area of a new, completed square. The side length of that square is the square's square root (50/60, or 5/6). This length is the unknown side length, plus 20/60 (1/3), so we just subtract 1/3 from 5/6 to find the unknown side length, which is 30/60 (1/2).

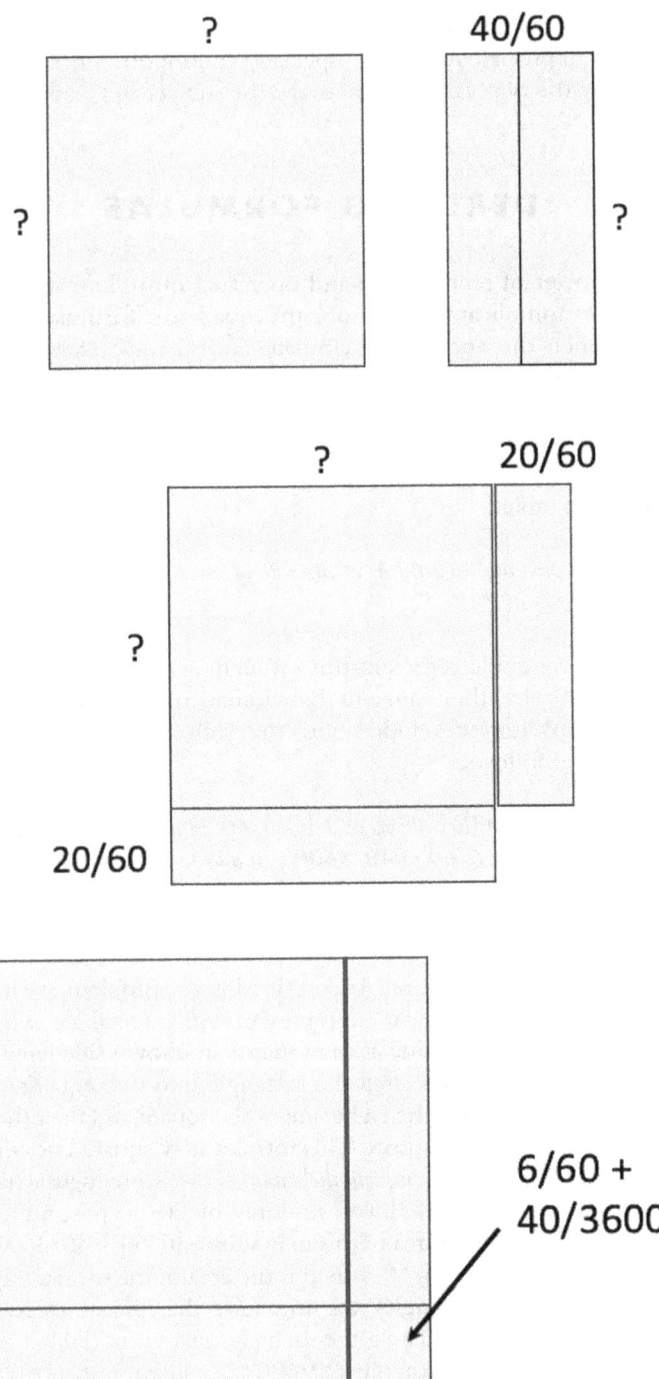

Figure 6.3 Completing the square.

We can generalize this solution—completing the square—to work with any quadratic equation (any equation of the form $ax^2 + bx + c = 0$). First, notice that we can transform the general form of a quadratic equation into the form $x^2 + \frac{b}{a}x = -\frac{c}{a}$, by subtracting c from each side of the equation and then dividing all terms by a.[11] So, we are in the same situation as before but with the value of b/a instead of 2/3 and the value of $-c/a$ instead of 35/60.

Reflection

Consider how you would relabel the side lengths and areas shown at the top of Figure 6.3, in order to generalize the Babylonian solution and derive the quadratic formula.

We have a square with side length x and a rectangle with width b/a and length x. By combining them, we have a rectangle with area $x^2 + \frac{b}{a}x$, and we know that area is also $-c/a$.[12] Now, we want to turn this rectangle into a square because, although many different rectangles can have a given area, only one square can have a given area. In other words, with the formula $A = l \times w$, lots of different lengths (l) and widths (w) can produce a specified area (A), but there is only one possibility when $l = w$.[13]

To turn the rectangle into a square we can split the original rectangle (with area $\frac{b}{a}x$) into two parts and place one on the right side of the original square and one on the bottom of it, as we did before. We can represent this split algebraically by dividing by 2 so that each new rectangle has an area of $\frac{b}{2a}x$. Together with the square, these rectangles nearly form a new square, with side length $x + \frac{b}{2a}$. However, a corner is missing.

The missing corner is a small square with side lengths $\frac{b}{2a}$. Algebraically, we would represent this area (and the two sweeps that form it) as $\left(\frac{b}{2a}\right)^2$. That means that the complete square would have an area of $-\frac{c}{a} + \left(\frac{b}{2a}\right)^2$, and so its side length would be the square root of that value.[14] But we already know its side length is $x + \frac{b}{2a}$, so we have $x + \frac{b}{2a} = \sqrt{-\frac{c}{a} + \left(\frac{b}{2a}\right)^2}$. Taking away the b/2a part of that length, we have $x = -\frac{b}{2a} + \sqrt{-\frac{c}{a} + \left(\frac{b}{2a}\right)^2}$. That formula might look familiar. With a little more manipulation and the inclusion of a negative root, we can show that it is the same as its more familiar form: $x = \left(-b \pm \sqrt{(b^2 - 4ac)}\right)/2a$. Thus, we have derived an algebraic formula—the quadratic formula—from the geometric process of completing a square, by using algebraic symbols as a proxy for our geometric actions.

As noted in Chapter 5, we can rely on geometric figures to represent coordinations of mental actions. In fact, ancient Babylonians used geometric figures to solve some quadratic equations, like the one at the start of this section. The ancient Greeks used geometric constructions, with a straight edge and a compass, to solve wider classes of quadratic equations. As such, they offloaded some of the mental labor of coordinating all the associated actions. Algebra offers an even more versatile tool for offloading those demands so that we can derive ever more abstract formulas.

ESTABLISHING IDENTITIES

Through symbolic manipulation of unknown quantities, we can generate equalities that hold regardless of the values of the quantities. For example, $(m + n)^3 = (m + n)(m + n)(m + n)$. If we associate the first two terms and distribute, we get $(m^2 + 2mn + n^2)(m + n)$, and then distributing once more, we get $m^3 + 3m^2n + 3mn^2 + n^3$. We could justify these manipulations geometrically or arithmetically—by referring to the underlying mental actions we could perform on squares or numbers—as we have in prior examples. The point here is that, by performing these manipulations on unknown quantities, m and n, we derive the following identity: $(m + n)^3 = m^3 + 3m^2n + 3mn^2 + n^3$. This identity will hold for all values of m and n on which we can perform these manipulations. Specifically, it will hold for any real numbers, m and n. Suppose for example, m = 10 and n = 1. The identity demonstrates that 11^3 equals 1331.

Reflection

What other interesting equalities can you generate from the identity? For example, can you use it to compute 9^3 or to find the difference of consecutive cubed numbers (e.g., $7^3 - 6^3$)?

You might recognize the numbers 1 3 3 1 from Pascal's/Pingala's triangle, mentioned in the introduction. Therein lies yet another connection we could justify by relating the actions used to produce Pingala's triangle to the actions we used to expand $(m + n)^3$. This connection also relates to probability and binomial expansion. For example, if you flip a coin eight times, you should expect to get all heads once (1), two heads and one tail three times (3), one head and two tails three times (3), and three tails once (1). In other words, it's three times as likely that you get two heads and one tail, or vice versa, as it is that you will get all heads or all tails.

Coincidentally, 1331 CE was also the year the Black Death began in Europe. Following a series of such calamities, the Renaissance began in Italy, and algebra played its part. In particular, del Ferro[15] used the identity $(m + n)^3 = m^3 + 3m^2n + 3mn^2 + n^3$ to solve cubic equations like $x^3 = 3x + 4$.

Generally, cubic equations are of the form $ax^3 + bx^2 + cx + d = 0$. A cubic formula would allow us to determine the values of x that solve the equation, in terms of the coefficients a, b, c, and d. Looking back at our derivation of the quadratic formula, from completing the square, you might imagine that a similar derivation (completing the cube?) could become very complicated, and you would be right. Algebra becomes especially useful in cases like this.

With a little algebraic manipulation, the identity $(m + n)^3 = m^3 + 3m^2n + 3mn^2 + n^3$ becomes $(m + n)^3 = 3mn(m + n) + m^3 + n^3$, and this identity holds for any real values of m and n. So, we can take the sum, $m + n$, to be the same as an unknown solution for a cubic equation. Specifically, let $m + n$ be a solution for the cubic equation $x^3 = 3x + 4$. In other words, let $m + n = x$ so that we have $(x)^3 = 3mn(x) + m^3 + n^3$. As a solution of the equation $x^3 = 3x + 4$, we know that x will also be a solution of $(x)^3 = 3mn(x) + m^3 + n^3$ when $3mn = 3$ and $m^3 + n^3 = 4$. So, the problem of finding a root to the cubic equation $x^3 = 3x + 4$ is reduced to finding a solution to the following system of equations: $3mn = 3$ and $m^3 + n^3 = 4$.

The first equation implies that $n = 1/m$. Substituting that value of n into the second equation we get $m^3 + (1/m)^3 = 4$, which is to say that $m^6 - 4m^3 + 1 = 0$. Although this sixth-degree polynomial looks more complicated, if we substitute $u = m^3$, we can reduce it to a quadratic equation.

Reflection

Perform the substitution suggested earlier and use the quadratic formula to find a value of u that solves the quadratic equation. Working backward through the substitutions, you should find a solution, x, for the original cubic equation: $x^3 = 3x + 4$.

Generalizing this process, 16th-century Italian mathematicians derived a cubic formula that could be used to solve any cubic equation, just like the quadratic formula solves any quadratic equation. They even derived a quartic formula for solving fourth-degree polynomials. These formulas relied on lots of algebraic manipulation. In Figure 6.3, we represented geometric transformations represented by algebraic manipulations, in service of deriving the quadratic formula. As we move to cubic and quartic formulas, the geometric transformations increase in both complexity and dimension so that we have to rely on algebraic manipulation more and more. The algebraic manipulations then serve as proxies for geometric transformations that we could neither visualize nor supervise. No doubt, Italian Renaissance mathematicians would have extended this algebraic program to the derivation of a quintic equation if possible. In Chapter 10, we consider the vanity of such efforts.

With modern technology, we might wonder why mathematicians spent so much time deriving formulas for finding solutions for polynomial equations. After all, we can readily graph polynomials of any degree and see the solutions (roots) as points where the graph crosses the x-axis. However, this argument raises two objections. First, unlike engineers, mathematicians concern themselves with exact values, which we can derive from a finite sequence of mental actions (often represented by algebraic manipulations). Second, graphing—which relies on the coordination of two covarying quantities (the subject of Chapter 9)—was not a thing in the 16th century. In fact, these Renaissance mathematicians didn't necessarily consider x as something that varies at all—only as an unknown value that they might determine through some process, such as the quadratic formula.

SUMMARY

Beginning with numerals, mathematical symbols provide a means for recording the products of our mental actions, so that we don't have to keep them in mind all at once. In terms of cognitive science, symbols provide a means for us to offload demands on working memory. This limited resource is responsible for coordinating mental actions in service of a goal, such as solving a mathematical task.[16] By offloading some of the demands to symbols, we free up working memory to deal with ever more challenging tasks.

As generalized arithmetic, algebra does even more. We can symbolize unknown quantities and operate on those symbols as if they were numerals. We can also represent geometric objects with algebraic symbols (e.g., x^2). Thus, algebraic manipulations serve as a proxy for numerical and geometric transformations. Although ancient Babylonian mathematicians derived a version of the quadratic formula geometrically, without algebra they could not manage the complexities (and extra dimensions) of deriving higher-degree formulas. Quickly following the invention of algebra (and the Renaissance), Italian mathematicians derived cubic and quartic formulas. Thus, algebra readily began to prove its power.

Nineteenth-century German mathematician Emmy Noether captured the sentiment as follows: "My methods [of algebra] are really methods of working and thinking; this is why they have crept in everywhere anonymously."[17] For algebra to become a method of thinking, the symbols must have meaning, which they derive from coordinations of our own mental actions. At that point, algebra becomes more than meaningless symbol manipulation; it becomes a tool for transcending the limitations of working memory.

Students generally have an intuitive sense for the power of symbols in aiding their thinking. This is why they often draw pictures or even invent their own notational devices when solving mathematical tasks.[18] Such invented symbols may be unconventional, but they hold personal meaning for the student, and conventional symbols come with their own challenges. For instance, the exponent -1 has different intended meanings in different mathematical contexts: whereas $f^{-1}(x)$ refers to an inverse function, x^{-1} refers to the reciprocal, $1/x$, and so it's no wonder students often confuse $\sin^{-1}(x)$ for $1/\sin(x)$ rather than the arcsine of x.[19] For symbols to have meaning, we must resolve such ambiguities. Then we might meaningfully operate on symbols but only when those operations serve as a proxies for actions we might carry out on the mathematical objects the symbols represent.

Activities

Activity 1: Try solving the following problem in your head, without writing anything down: "Given a string that is 5/9 of a meter long, can you determine the length of 1 meter?" Reflecting on the mental actions you performed, draw a unit transformational graph, like the one in Figure 6.1, to represent your coordination of actions in solving the task. Try this again for more and more complicated tasks, like the following task (from Activity 2, in Chapter 4): "Using only a one-third measuring cup, how can you measure off 3/4 of a cup of flour?" At what point do you need to rely on symbols or drawings to solve the task?

Activity 2: Draw a rectangle with a length of a + b and a width of c + d. Use the area of the rectangle to justify distribution: $(a + b) \times (c + d) = ac + ad + bc + bd$.

Activity 3: In the case of a square, with side lengths (m + n), we get that $(m + n)^2 = m^2 + 2mn + n^2$. Generalize this geometric connection to represent the volume of a cube as $(m + n)^3 = m^3 + 3m^2n + 3mn^2 + n^3$. It's difficult to visualize what happens in dimensions four and higher, but you might generalize the process and see that it's the same process that generates rows in Pingala's triangle.

Activity 4: You might also algebraically generalize the Pythagorean theorem to derive a distance formula for three-dimensional space (and to an equation for n-dimensional spheres). Starting with a rectangular prism (a box) with width a, length b, and depth c, you can apply the Pythagorean theorem twice to find the diagonal that connects its opposite vertices (corners). Could you do something similar for a four-dimensional box?

NOTES

1. In referring to numerals, von Glasersfeld (1995) describes "the pointing power of symbols" (p. 173), and the title of this chapter matches one from his book.

2. This task is taken from Hackenberg and Tillema (2009, p. 10) whose research demonstrated tight connections between children's multiplicative reasoning, fractions knowledge, and their ability to coordinate units in general.

3.   Philipp and Schappelle (1999) raise this issue and practical ways to address it in yet another valuable article from *The Mathematics Teacher*.

4.   These "unit transformation graphs" were developed by Norton, Ulrich, and Kerrigan (in review).

5.   Note that the word, *algebra*, comes from the Arabic word *al-jabr*, which means "bone-setting." When al-Khwarizmi first invented algebra, he referred to it as the art of reunion and reduction, which included combining like terms into a single unit, akin to resetting a broken bone (Burton, 2007).

6.   The Philipp and Schappelle (1999) article, mentioned previously, discusses this idea too.

7.   Hackenberg and Lee (2015) investigated this and other connections between students' abilities to coordinate units and their development of algebraic reasoning.

8.   This problem appears as Problem #23 in the Rhind Papyrus, written by the scribe Ahmes around 1650 BCE (Burton, 2007).

9.   Note that the Egyptians did not recognize nonunit fractions, so they wrote all fractions as an amalgamation of unit fractions (Burton, 2007). It's as if they wanted to take chunks out of the fraction, one unit at a time, starting from the largest unit fraction, 1/2. This method of breaking down numbers into chunks fits their general approach to arithmetic, including their methods for multiplication, division, and (as we have just seen) solving for unknowns.

10.   The Babylonian problem and its solution are presented in Burton (2007, p. 65). Note that the problem includes 60ths because the ancient Babylonians used a base-60 number system.

11.   Here, we assume a ≠ 0; otherwise, we don't really have a quadratic equation.

12.   Note that, while −c/a has a negative sign, its value can be positive. For example, in the previous case, we had c = −35/36 and a = 1, so −c/a was the positive value 35/60.

13.   This is the reason "completing the square" works as a method for solving quadratic equations but "completing a rectangle" would not. Factoring works as a method for finding roots of quadratic equations only when the product of the two factors is equal to zero because, in that case, we know at least one of the factors must be 0.

14.   Note that there is a positive root and a negative root, but as a length, we only consider the positive root as ancient Babylonian mathematicians would have done.

15.   del Ferro had a method, like the one described here, for solving all cubic equations of the form $x^3 + hx = k$, where h and k are positive numbers. As it turns out, by using a substitution to eliminate the x^2 term, all cubic equations can be reduced to this form (Burton, 2007)

16.   Agostino, Johnson, and Pascual-Leone (2010, p. 62) describe the role of working memory as "holding information in an active state and manipulating it until a goal is reached."

17.   This quote comes from a letter to Helmut Hasse, in 1931, as cited by Rowe and Koreuber (2020, p. 37). The authors suggest the quote also alludes to the fact that much of Noether's work in algebra went unrecognized ("crept up anonymously") in the work of other mathematicians, including some of her early supporters, like David Hilbert.

18.   For example, Tillema and Hackenberg (2011) describe ways that students can develop systems of notation for supporting their reasoning with fractions.

19.   Rina Zazkis (2016) investigated this sort of ambiguity with preservice teachers and found it related to their understandings of inverses within algebraic groups (as studied in Abstract Algebra courses).

REFERENCES

Agostino, A., Johnson, J., & Pascual-Leone, J. (2010). Executive functions underlying multiplicative reasoning: Problem type matters. *Journal of Experimental Child Psychology*, *105*(4), 286–305.

Burton, D. M. (2007). *The history of mathematics: An introduction.* New York: McGraw-Hill.

Hackenberg, A. J., & Lee, M. Y. (2015). Relationships between students' fractional knowledge and equation writing. *Journal for Research in Mathematics Education*, *46*(2), 196–243.

Hackenberg, A. J., & Tillema, E. S. (2009). Students' whole number multiplicative concepts: A critical constructive resource for fraction composition schemes. *Journal of mathematical behavior*, *28*(1), 1–18.

Norton, A., Ulrich, C., & Kerrigan, S. (in review). Unit transformation graphs: Modeling students' mathematics in meeting the cognitive demands of fractions multiplication tasks. *Journal for Research in Mathematics Education*.

Philipp, R. A., & Schappelle, B. P. (1999). Algebra as generalized arithmetic: Starting with the known for a change. *Mathematics Teacher*, *92*(4), 310.

Rowe, D. E., & Koreuber, M. (2020). *Proving it her way: Emmy Noether, a life in mathematics.* Cham, Switzerland: Springer.

Tillema, E., & Hackenberg, A. (2011). Developing systems of notation as a trace of reasoning. *For the Learning of Mathematics*, *31*(3), 29–35.

von Glasersfeld, E. (1995). A constructivist approach to teaching. In L. P. Steffe & J. Gale (Eds.), *Constructivism in education* (pp. 3–15). Hillsdale, NJ: Erlbaum.

Zazkis, R. (2016). A curious case of superscript (-1): Prospective secondary mathematics teachers explain. *Journal of Mathematical Behavior*, *43*, 98–110.

7

Seeing Angles From a Different Angle

I n previous chapters, we have referred to angle measures in degrees. We say, "There are 360 degrees in a circle," and have the Babylonians to thank for that. In Chapter 1, we noted that the Babylonians chose a base-60 number system because 60 is a highly divisible number; it is the smallest number divisible by 2, 3, 4, and 5 (and you get 6 for free). This divisibility is also useful for partitioning angles within a circle, whose circumference can be approximated by the perimeter of a regular hexagon (see Figure 7.1).

The regular hexagon is made up of six equilateral triangles. By Babylonian accounting, equilateral triangles have interior angles of 60 degrees each, so the six angles that meet at the center of the circle add to 360 degrees (a number that also approximates the number of days in a year[1]). We have grown accustomed to this custom, and it is useful, but like part–whole conceptions of fractions (see Chapter 4), it is also limited. A more powerful conception of angle measure relies on a new unit for measuring: the radian.

A radian is an arc length (a distance along the circumference of the circle) measured in radii, hence the name radian. Figure 7.1 shows that the circumference of a circle measures a little more than six radians—2π radians, to be exact. The close proximity between 6 and 2π (about 6.28) shows how closely the perimeter of the hexagon approximates the circumference of the circle. Suppose that the circle shown in Figure 7.1 has a radius of 1. Then the sides of the regular hexagon

DOI: 10.4324/9781003181729-8

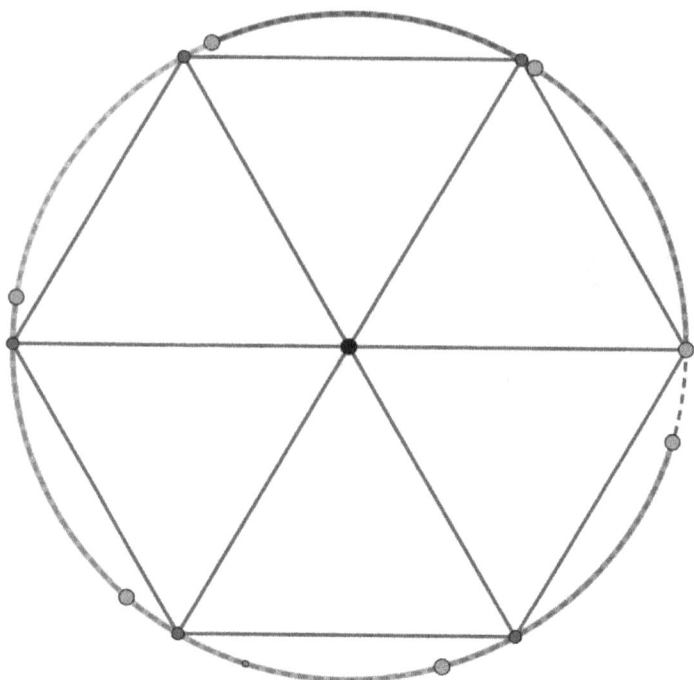

Figure 7.1 The circle, the hexagon, and the radian.

would also be 1 unit long, and so would the length of a radian. However, the radian is measured along the curved arc of the circle, so it doesn't quite span the side of the hexagon. We see this difference add up: as we go around the circle, counterclockwise, we're left with a gap (the dashed line in Figure 7.1), which measures about 0.28 units (radians).

π is defined as the ratio of a circle's circumference to its diameter, which makes 2π the ratio between the circumference and radius. To see that this ratio is the same for any circle, just consider the fact that all circles are similar to each other (they only differ in their size, as determined by their radius), and when a circle is dilated, all lengths associated with it are scaled by the same factor. In particular, the dilation will scale the radius and circumference of the circle by the same factor so that their ratio will be the same as it was before the dilation. Here again, it's like fractions, where one fraction has the same measure as another if we scale the number of iterations and partitions by the same factor (as investigated in Chapter 4). For example, 2/3 and 4/6 are equivalent because we have doubled the number of iterations and partitions.[2] In partitioning, the size of the unit fraction is transformed from 1/3 to 1/6, but in iterating, the measure of the fraction measured in those units has changed proportionally, from 2 to 4.

As with unit fractions, we can think about the radian as a unit of measure. It measures arc lengths along the circle's circumference, and at the same time, measures the central angle subtended by that arc (i.e., the arc spans the openness of the angle). This measure is independent of the particular size of the circle because, whatever its length, the radius is treated as a unit of 1. Developing this understanding of angle measure is no small feat, but it generates powerful tools for understanding geometry and trigonometry alike.

DEVELOPING CONCEPTS OF ANGLE

What are we measuring when we measure an angle? When we measure length, there is something tangible to reference, such as a line segment. When we measure the distance between two points, we can consider the line segment swept out as we shift our attention from one point to the other. The line segment serves as a proxy for the shift we are measuring. It leaves a tangible trace and a figurative referent for the mental action of shifting. Measuring angles is different because the mental action is different—a rotation rather than a shift—and because there is no readily available trace of that mental action.

Studies in neuroscience suggest that the mental action of rotating mirrors the physical action we would carry out to perform the rotation on a physical object.[3] In particular, mental rotations take longer to perform when the angle of rotation is larger, just as the physical action would take longer to perform. Research has demonstrated success in leveraging physical embodiments of rotation to teach fourth-grade students about angle measure.[4]

High school textbooks generally define angles with a figurative referent. For instance, they might say that an angle is two rays meeting at a point (the vertex of the angle). Understandably, students will then try to measure the angle by measuring the distance between the two rays, which grows larger when measured farther from their common vertex. What we need to measure is not a distance but the extent of our own mental action in rotating one ray to the other. Radians provide particularly useful measures of angle because they leave a trace of the rotation and its extent, as an arc length or an angular sweep.

Researchers have studied the conceptual development of angle measure among preservice secondary school teachers, using figures like the one shown in Figure 7.2. Referring to that figure, we might ask the following questions:

1. If the radius of the smaller circle is 2 and the radius of the larger circle is 5, and if angle a measures 1 radian, what are the lengths of the two arcs?
2. If the radius of the larger circle is 3 and the arc is π units long, what is the measure of angle a?
3. In general, can you find a formula that relates the radius of a circle, the measure of angle a, and the arc length that subtends that angle?

Reflection

How might you answer the questions listed earlier?

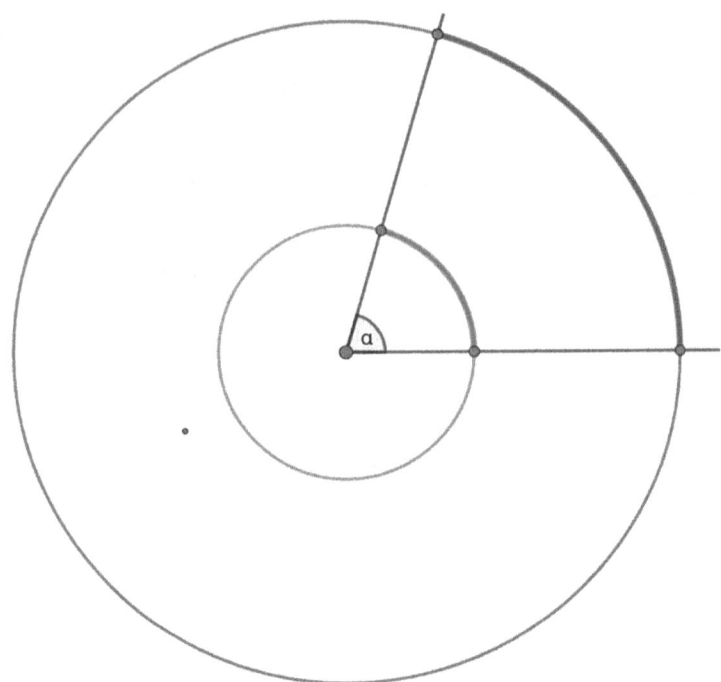

Figure 7.2 Angle task.

Source: Adapted from Moore (2013).

Productive struggle with such questions may help students understand angle measures as "fractional amounts of a circle's circumference."[5] In turn, that understanding, and its connection to students' own mental actions of rotating, unlocks their potential to make new connections. So far, we have considered connections between central angles, arcs lengths, and the radius of a circle. In the next section, we consider a theorem that relates central angles to inscribed angles and an even simpler proof of Thales' theorem (first proved in Chapter 3).

TWO TWO-CHORDS THEOREMS

Suppose two chords intersect inside of a circle with radius 1 (i.e., the unit circle), breaking its circumference into four parts, as shown on the left side of Figure 7.3. Is there anything we can say about the four angles that are formed?

You might say that angles a and c must be congruent (likewise for angles b and d), and you would be right, though I'd be curious about the argument you'd use. After all, the arc lengths that subtend these pairs of angles are not congruent. Is it a rule you've learned about vertical angles? Can you justify this rule on the basis

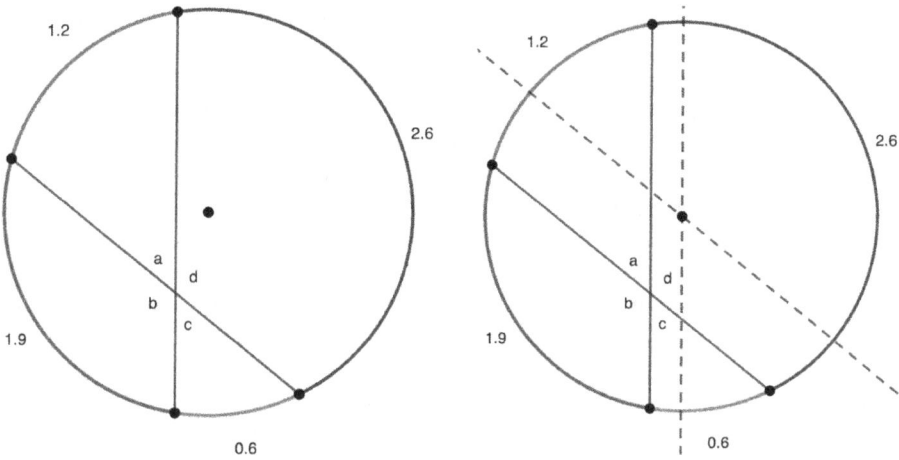

Figure 7.3 Two Chords Theorem.

of your own mental rotations? Once you've done that, maybe you can determine their angle measure. Is it 0.6 radians, 1.2 radians, or something in between?

Reflection

Once more, take a moment to answer the questions raised earlier before moving on. The right side of Figure 7.3 provides figurative cues that may or may not need.

By the two chords theorem, the vertical angles, a and c, each measure (0.6 + 1.2)/2. In general, each pair of vertical angles will measure the average of the two arc lengths that subtend them. We can prove this theorem transformationally by extending the chords into lines and translating them so that they intersect at the center of the circle (see the right side of Figure 7.3). In doing so, we don't change the angles at which the two chords meet, but we do change their position so that their measure will correspond with arc lengths on the circle. In other words, as measures of rotations, the angle measures (though not the arc lengths) are invariant under the mental action of translating the two chords.

Notice that our translation of the chords does change the lengths of the four arcs. However, the length taken from one arc is added to the opposite arc so that the pairs of opposite arcs balance out to their average arc length, as the two chords theorem states. It may be easier to see this if we perform that translation in two steps: first translating one chord to its new, parallel position; then doing the same for the other chord. For example, when the vertical chord in Figure 7.3 is translated to its new position (along the vertical dashed line), a piece of the arc on top is added and a same-sized piece of arc on the bottom is taken away. We know that the two pieces are the same size because of the symmetry of the circle (e.g., rotating the circle halfway, by π, about its center, and reflecting over the vertical line through this center, will map these two pieces to each other).

Here again, the figures (e.g., the intersecting chords and the arc lengths) help us keep track of our mental actions and how they affect each other. In particular, the angles are defined by mental rotations, but the intersecting chords and arc lengths mark the extent of those rotations. While the intersecting chords are not affected by the new action of translating, the arc lengths are. Once the chords intersect at the center of the circle, the arc lengths will correspond to the same angle measure—measured in radians. In the example under consideration, one pair of arc lengths becomes 0.9 radians, and the other pair becomes 2.25 radians.

With the two chords theorem, we can use arc lengths to measure angles even when the angles aren't marked at the center of the circle (i.e., central angles): we just average the two arc lengths across from the vertical angles. A special case arises when the two chords meet on the circumference of the circle. In that case, the chords form an inscribed angle. What does the two chords theorem say about the measure of an inscribed angle? For example, consider the inscribed angle at point P, in Figure 7.4.

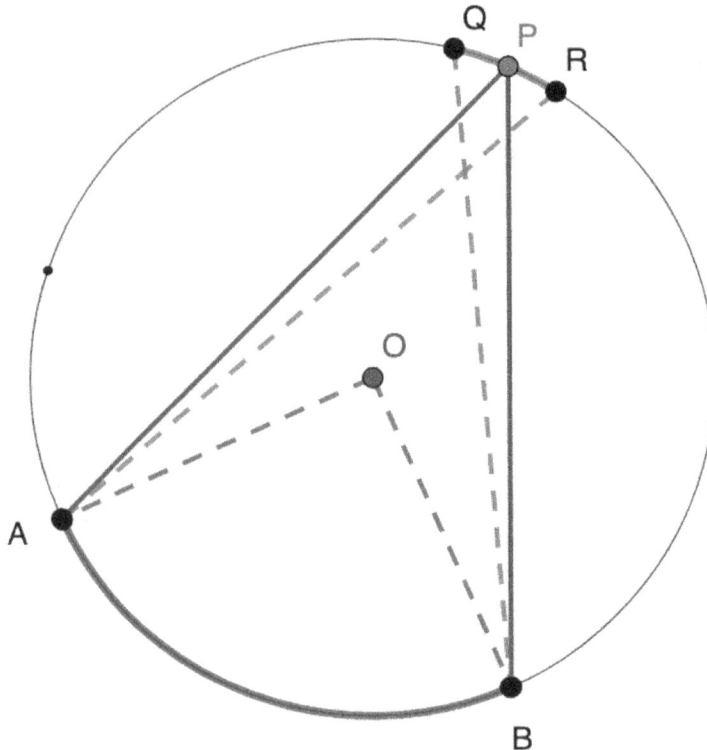

Figure 7.4 Inscribed angles, central angles, and the two chords theorem.

Note that segments AP and BP form two chords of the circle, centered at O. So the two chords theorem suggests that angle APB is the average of two arc lengths. One of these arc lengths is the one from point A to point B (in bold), but there is no arc length opposite it. Should we consider that arc length to be zero? If so, then the average of the two arc lengths would be half of the central angle AOC. In other words, any inscribed angle would measure half the arc length that subtends it.[6] In this case, the angle at P would measure half of the arc length from A to B. We'll return to this investigation, with reference to chords AR and BQ, in an activity at the end of the chapter, along with a new proof of Thales' theorem (first proved in Chapter 2).

Figure 7.5, with its two intersecting chords (AC and BD), resembles Figure 7.4, but this time we're focused on the lengths of the four segments they form, rather than the measures of the four angles they form. What can we determine about the lengths of a, b, c, and d?

If you need a hint, consider drawing in a segment from A to D and a segment from B to C, and then consider the angles formed, relying on the inscribed angles theorem. For instance, angles PBC and PAD are subtended by the same arc, CD, so they must be congruent. In fact, you should find that

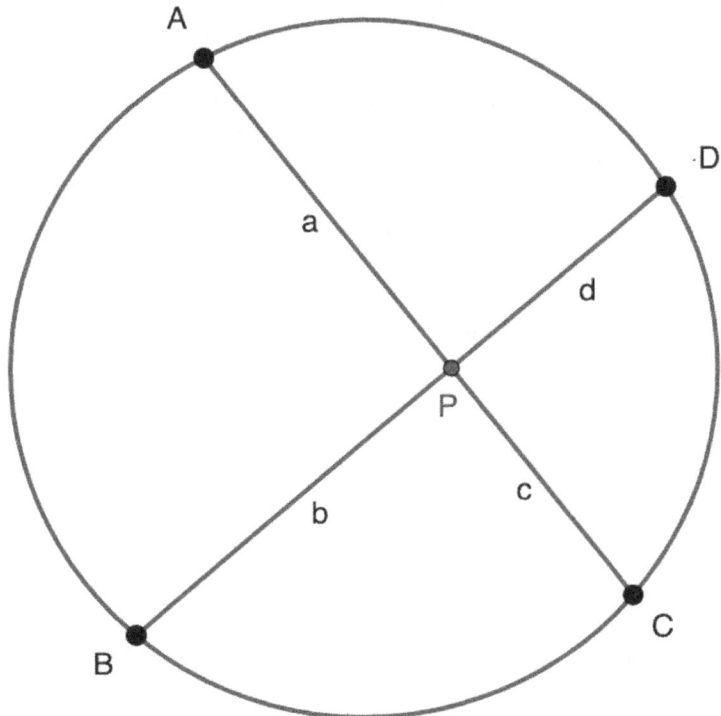

Figure 7.5 Intersecting chords.

triangles PAD and PBC are similar triangles, so their respective sides are proportional; namely, a is to b as d is to c, and thus, a × c = b × d. This finding gives us a second two-chords theorem, called the Intersecting Chords Theorem. The two two-chords theorems are instrumental in proofs of all kinds of advanced geometry theorems, including one about "mystic hexagons," as we will see in the next section.

MYSTIC HEXAGONS

We have been using the word *subtend* to describe arc lengths that span the openness of an angle, and these arc lengths provide us with its measure, in radians. However, line segments can also subtend an angle. Might they also provide good measures? For example, consider the circle on the left side of Figure 7.6.

By the inscribed angles theorem, we know that the openness of angle APB is measured by half of the arc that subtends it (arc AB). What about segment ST? It subtends the same angle, so what might its length tell us about the measure of that angle? As it turns out, not very much, unless we know how that line segment is positioned within the circle. In particular, if the subtending segment were oriented differently, or moved further away from point P, it might need to be longer in order to subtend the angle. If we moved point P around the circle, the segment would need to transform accordingly (though arc AB would not).

Already, we can sense the complexity in measuring angles with segments. There are numerous mental actions to coordinate here: as point P rotates along the circumference of the circle, the subtending segment rotates, dilates, and/or shifts accordingly. Just consider how the position of the subtending segment (originally ST, on the left side of Figure 7.6) changes as we move P to P'. With its length

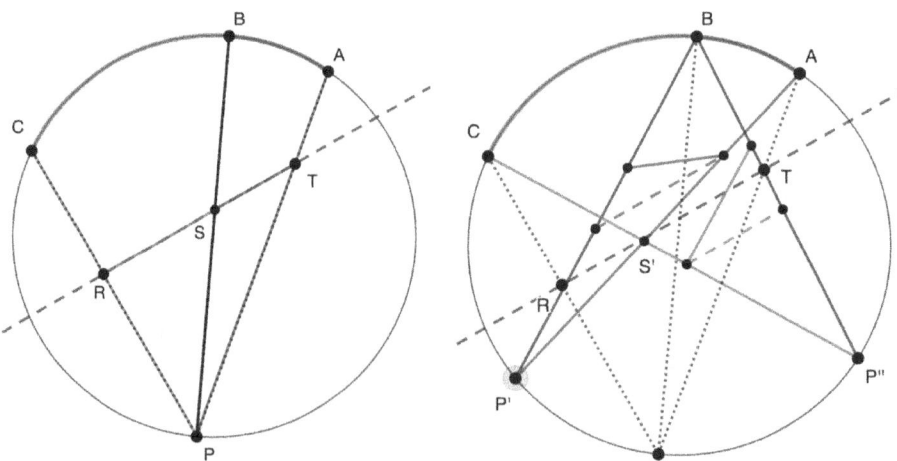

Figure 7.6 Subtended angles.

preserved, the segment rotates so that it still subtends angle APB (now AP'B). The dashed segment helps show the angle of rotation. Likewise, as we move P to P″, the segment that subtends angle BPC (segment RS) must rotate so that its length is preserved and it continues to subtend the angle.

Reflection

Imagine the segment rotating as P moves along the circumference of the circle to P'? Can you determine how much it rotates?

Something "mystical" happens when the vertex of angle APB is moved to P' and the vertex of BPC is moved to P″ (as shown on the right side of Figure 7.6). Notice that P'B passes through R, P″B passes through T, and that P'B and P″C intersect at S', along the same (dashed) line. Seen from a different angle, angles APB and BPC have switched sides, but their sum is still subtended by ST. This phenomenon is known as Pascal's theorem.[7]

Recall Pascal's triangle from the Introduction. Recall, too, that other mathematicians, such as Pingala, had studied it generations before. So, it's ironic that Pascal's name is so closely associated with that triangle, especially when his original contributions to mathematics, science, philosophy, and theology were so numerous and profound.[8] He proposed his theorem about "mystic hexagons" when he was only 16 years old.

The theorem states that, for any hexagon inscribed in a circle, its three pairs of opposite sides will intersect along the same line. Figure 7.7 illustrates the case in which these intersections occur inside of the circle, making it easier to visualize. The hexagon may look strange (and yet familiar; see the right side of Figure 7.6), but you can see a six-sided closed figure formed by following its six vertices in this order: A–B–C–D–E–F and then back to A.

The order of the vertices indicates the following three pairs of opposite sides: AB and DE, BC and EF, and CD and FA. In Figure 7.7, the first two pairs intersect at R and T, respectively. These two points define a (dashed) line. Pascal's theorem stipulates that the remaining pair of sides (CD and FA) intersect on this same line. There are several proofs of Pascal's theorem—one of which we'll examine in an activity at the end of the chapter—but here we outline an argument that relies on mental transformations within the figure.

Returning to the right side of Figure 7.6, we might transform the subtending segments back onto the segment RT. The original segments, RS and ST, will have changed lengths and switched sides, relative to their positions on the left side of Figure 7.6, but their sum will remain the same. In particular, recall that the subtending segments on the right side of Figure 7.6 have the same length as the subtending segments on the left side of Figure 7.6, but they have been rotated. We could rotate them back so that they are parallel to the line through RT. In so doing, we would map onto the two dashed segments.

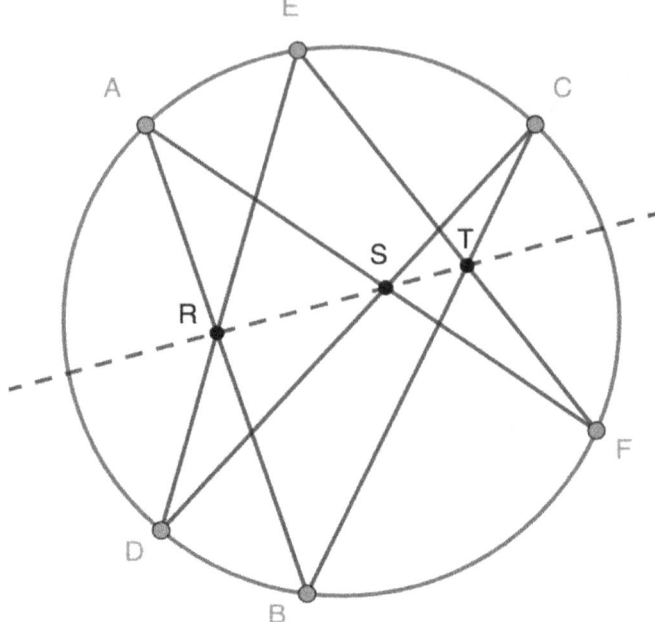

Figure 7.7 Pascal's mystic hexagon.

Reflection

By what ratio will the lengths of the two segments change (dilate) as you rotate them to the two dashed segments?

Then, we could dilate these dashed segments, with respect to points P′ and P″, respectively, so that they are mapped onto segment RS′ and S′T, respectively. Proportions given by the two chords theorem would tell us how the lengths would change in these dilations. We could then show that the sum of the two transformed segments is the length of RT, as before. Thus, we would show that AP′ and CP″ intersect (at S) along the line through R and T.

The proof outlined earlier illustrates two characteristics of mathematics: that mathematical phenomena arise from transformations that we perform for ourselves, and that these transformations can become increasingly complex. Complexities owe to the sheer number of transformations we perform and attempt to coordinate, which explains why we often rely on figures and algebraic notation to keep track of them all! They also owe to new kinds of transformations (mental actions) that we learn to perform. In the case of Pascal's theorem, we can extend those transformations to include projections, which generalize the theorem to include hexagons inscribed in any conic section (e.g., ellipses and hyperbolas).

REVISITING PROJECTIVE GEOMETRY

Chapter 3 introduced the mental action of projection using the familiar experience of projecting an image on a screen. More formally, we might think about a projection as a transformation from a particular point of view. Rays emanate from a single point, P, through space and onto a plane in that space. Any figure those rays pass through gets mapped onto a corresponding figure in the plane. In Chapter 3, we considered the example of a square getting mapped onto a trapezoid. Here, we consider the projection of a circle onto an ellipse (see Figure 7.8).

Projections do not preserve lengths or angles, but they do preserve lines and intersections. Thus, the hexagon inscribed in the circle gets projected onto a hexagon inscribed in the ellipse, and the three points of intersection of its three pairs of opposite sides will remain colinear. Consider for a moment the consequences of this transformation: it suggests that any hexagon inscribed in any ellipse will have the same property!

Reflection

Try drawing an ellipse, and choose any six points on it. What happens when you connect the points to form a self-intersecting hexagon (like the one shown in Figure 7.8)?

Projective geometry demonstrates the power of this new transformation (projection), which allows us to generalize results from Euclidean geometry. We can also use it to make Euclidean proofs simpler. For example, we might begin by proving Pascal's theorem for special cases, such as when the hexagon inscribed

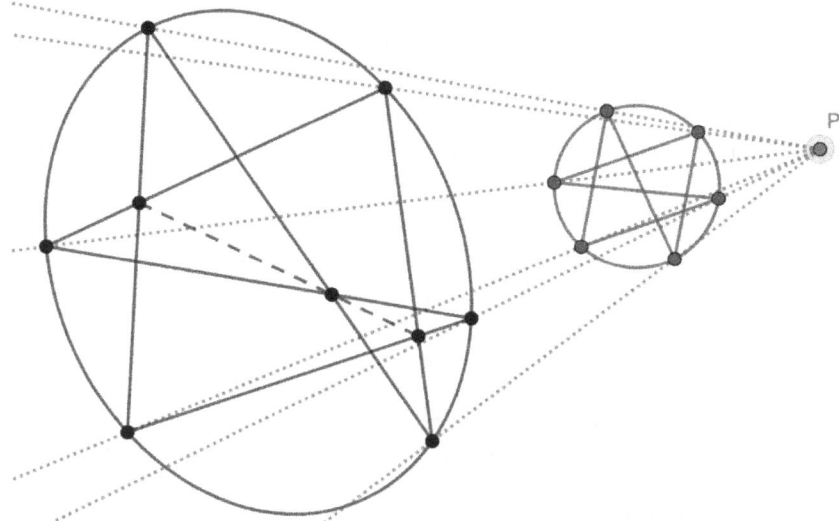

Figure 7.8 Generalizing Pascal's theorem via projection.

in the circle has two pairs of parallel sides. Then we can generalize that proof by projecting the circle onto an ellipse and then back onto circles with hexagons that don't necessarily have parallel sides.[9]

SUMMARY

Despite the merits of the base-60 number system and its contribution to measuring angles in degrees, the arguments made in this chapter should demonstrate the utility of measuring angles in radians instead. Specifically, arc lengths provide us with a physical referent for measuring the extent of our mental rotations so that, from a single pair of chords, we can determine angle measures, triangle similarities, and related proportions. In that regard, arc lengths offer a unique perspective on angle. When we consider other lengths that subtend an angle, the situation becomes complicated quite quickly! Nevertheless, with the aid of drawings and geometric constructions, we can keep track of them. In the process, we should not lose sight of the essence of mathematics. The transformations themselves, and not the figurative referents, define mathematical objects and their relationships.[10]

Activities

Activity 1. Returning to Figure 7.4, how would you explain that angle APB is half of angle AOB? If angle AOB were enlarged into a straight angle so that AB formed the diameter of the circle, what would be the measure of APB then? How does this relate to Thales' theorem from Chapter 3?

Activity 2. Suppose a chord intersects the diameter of the circle at a right angle. What can be said about the lengths of the four segments that are formed?

Activity 3. If an arc subtends an angle inscribed in a circle and the remaining circumference of the circle subtends another angle inscribed in that same circle, how are the angles related?

Activity 4. Pascal's theorem applies to convex hexagons as well as self-intersecting hexagons. Consider the convex hexagon inscribed in the solid circle, as shown in Figure 7.9. Its pairs of opposite sides, when extended, intersect at R, S, and T. In one proof of Pascal's theorem, Yzeren had the insight to construct a second (dashed) circle passing through S and two of the vertices of the hexagon.[11] He then proved that R must be colinear with S and T by showing that various angles in the figure are congruent and that the two bold and dashed triangles in Figure 7.9 are similar, with corresponding sides parallel. Thus, a linear projection from T projects S onto R, so R, S, and T are collinear.

Use inscribed angles to identify and label as many congruent pairs of angles as you can.

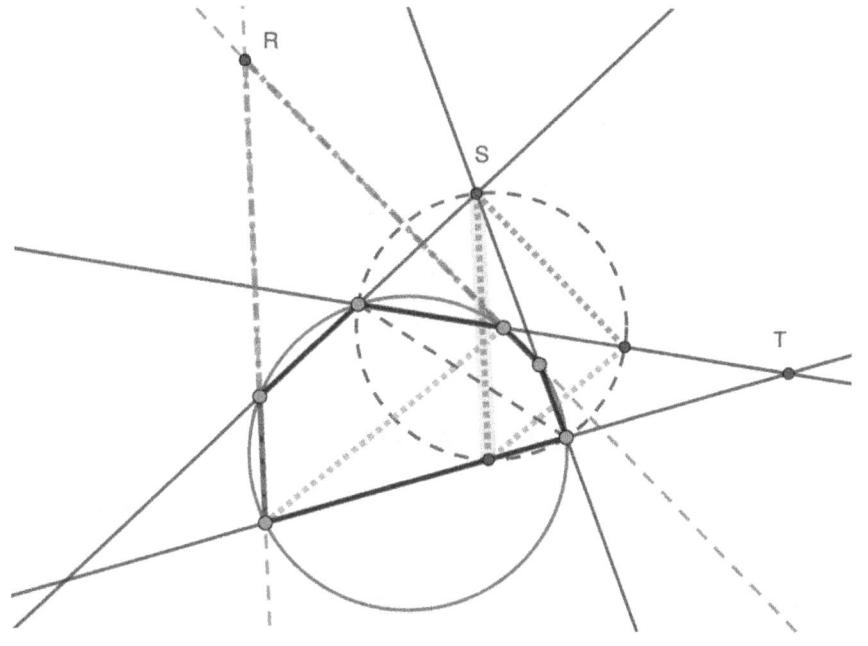

Figure 7.9 Yzeren's proof of Pascal's theorem

Source: Yzeren (1993).

Hint: Activity 3 could be useful.

NOTES

1.   See Matos (1990) for more on "the historical development of the concept of angle" and its relation to time and divisibility.

2.   Formally, we say that fractions form equivalence classes, where a single fraction represents an entire class of equivalent fractions. Likewise, radians form an equivalence class of arc lengths relative to radii (Moore, 2013). If an angle measures 1 radian, this means it spans an arc length equal to radius of the circle on which the arc lies.

3.   In a meta-analysis of several neuroscience studies, Zacks (2008) concluded that, at least in some situations, "mental rotation depends on motor simulation" (p. 1).

4.   In a study conducted by Clements and Burns (2000), students learned about angles by turning their bodies. They began to understand angle measure by using benchmarks, such as turning to face left, marking a 90-degree turn.

5.   This quote comes from Liang and Castillo-Garsow (2020, p. 53), who conducted a study of undergraduate students as they investigated the relationship between central angles and inscribed angles, similar to the investigation presented in the next section.

6.   The Inscribed Angle Theorem (also known as the Star Trek Lemma, for obvious reasons) appears as Proposition 20 in Book III of Euclid's *Elements*: https://mathcs.clarku.edu/~djoyce/elements/bookIII/bookIII.html. The Intersecting Chords Theorem (discussed next) appears as Proposition 35 in the same book.

7.   This theorem has been proved several times, in several different ways. For example, we will examine Yzeren's (1993) proof in the activities at the end of this chapter.

8.   Regarding the latter, Pascal wrote "the wager" as an argument for Christianity involving probability and expected values (Pascal, 1966/1670).

9.   In the *College Mathematics Journal*, Augros (2012) took advantage of projections to do just that.

10.   Portnoy, Grundmeier, and Graham (2006) report on a study with college geometry students demonstrating how easily we can lose sight of this essential character of mathematics.

11.   Yzeren's (1993) motivation for constructing the dashed circle is anything but obvious, but that insight renders the rest of the proof accessible with the use of the inscribed angles theorem and arguments about similar triangles.

REFERENCES

Augros, M. (2012). Rediscovering Pascal's mystic hexagon. *The College Mathematics Journal, 43*(3), 194–202.

Clements, D. H., & Burns, B. A. (2000). Students' development of strategies for turn and angle measure. *Educational Studies in Mathematics, 41*(1), 31–45.

Liang, B., & Castillo-Garsow, C. (2020). Undergraduate students' meanings for central angle and inscribed angle. *The Mathematics Educator, 29*(1).

Matos, J. (1990). The historical development of the concept of angle. *The Mathematics Educator, 1*(1). Retrieved from https://openjournals.libs.uga.edu/tme/article/view/1745

Moore, K. C. (2013). Making sense by measuring arcs: A teaching experiment in angle measure. *Educational Studies in Mathematics, 83*(2), 225–245.

Pascal, B. (1966). *Pensées* (A. J. Krailsheimer, Trans.). London: Penguin (Original work published in 1670).

Portnoy, N., Grundmeier, T. A., & Graham, K. J. (2006). Students' understanding of mathematical objects in the context of transformational geometry: Implications for constructing and understanding proofs. *The Journal of Mathematical Behavior, 25*(3), 196–207.

Yzeren, J. (1993). A simple proof of Pascal's hexagon theorem. *American Mathematical Monthly, 100*(10), 930–931.

Zacks, J. M. (2008). Neuroimaging studies of mental rotation: A meta-analysis and review. *Journal of Cognitive Neuroscience, 20*(1), 1–19.

8

The Geometry of Numbers

Mathematics has been defined as the study of space and number, but what unifies space and number into a single subject? For the Greeks, space was the medium for constructing number. Beginning with a line segment of length 1, the Greeks would swing circles, extend lines through pairs of points, and create points of intersection to construct new line segments, which represented numbers as lengths. We began to investigate this idea in Chapter 5 and return to it in Chapter 10. Here, we rely on geometry to extend numbers in different directions and into new dimensions, including one that we might call imaginary. These directed quantities enable us to reconsider numbers as vectors, with both length (magnitude) and direction. This extension of number into higher dimensions is the basis for linear algebra, known in ancient China long before the time of Carl Gauss and "Gaussian elimination" (row reduction).

EXTENDING NUMBERS TO VECTORS

In ancient India and China, negative numbers were accepted early on, as additive inverses to positive numbers. Seventh-century Indian mathematician Brahmagupta

DOI: 10.4324/9781003181729-9

even formulated rules for multiplying with negative numbers (e.g., a negative times a negative is a positive).[1] Because ancient Greek mathematicians considered numbers as lengths, they did not accept negative numbers as numbers at all. They rejected negative solutions as false and considered problems with only negative solutions absurd.[2] Negative numbers were as imaginary as imaginary numbers, and both appeared unintentionally, through algebra.

Today, we accept negative numbers as directed quantities.[3] What changed was the result of a simple idea: a change in direction, a reflection. We can think of negative numbers as positive numbers, reflected over the point 0 and extending in the opposite direction. At the flip of a switch, false solutions to absurd equations become mathematical objects composed of that reflection and the mental actions that define positive numbers.

Introducing new numbers into a number system introduces new problems to solve because combinations of numbers in the number system are also contained within the number system (closure). We have two principal operations for combining numbers: addition, which continues the iteration of units (as in the example of 2 + 2 = 4 discussed in Chapter 1), and multiplication, which transforms units (discussed in Chapters 1 and 4). Brahmagupta figured out rules for incorporating negative numbers under both operations. Here, we consider a geometric approach that explicitly relies on the mental action of reflection (the subject of Chapter 2).

After reflecting the positive numbers on the number line over the point 0, we have numbers extending in both directions—positive and negative—from 0 (see Figure 8.1). With the inclusion of negative numbers, numbers on the number line become endowed with both magnitude and direction. Thus, we can think about them as vectors, and as vectors, we can add them and transform them geometrically. In fact, we can think of the reflection over 0 as a transformation represented by the 1-by-1 matrix, [−1].

Like whole number multiplication, matrix multiplication transforms units. Referring to Chapter 1, we can think of the matrix as the multiplicand and the vector as the multiplier. When multiplying a vector by a matrix, the vector represents the number(s) of iterations of a unit (or units), and the matrix describes the transformation of units. For example, [−1]<5> represents the transformation of the vector <5> into the vector <−5> when multiplied by the matrix [−1]. Five iterations of the unit vector <1> are transformed into five iterations of the unit vector <−1>, or five units of <−1>, which yields the vector <−5>. The matrix [−1] has transformed the unit of <1> into a unit of <−1>. The geometric effect is that the vector <5> has been reflected over <0> to the vector <−5>.

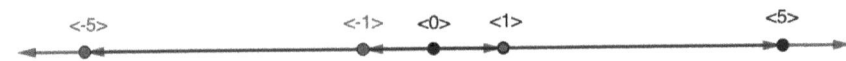

Figure 8.1 Number line with directed numbers.

Reflection

By considering what matrix multiplication does to individual vectors, we can imagine how a matrix would geometrically transform the entire number line. For example, [−1] reflects the entire number line over 0. What geometric transformations of the number line do the following matrices represent: [7], [1/2], [−3], and [0]?

Like whole number addition, vector addition continues the iteration of units. Positive vectors build from a unit of <1>, and their sums combine magnitudes in that positive direction. Likewise, negative numbers build from a unit of <−1>, and their sums combine magnitudes in the negative direction. Vectors with the same magnitude and opposite directions are additive inverses because when we add them, the iterations of their units balance (or cancel) out to 0, resulting in the zero vector, <0>. In other words, iterating by units of <−1> is the same as reversing, or undoing, iterations of <1> and vice versa. As with whole numbers—and as the inverse operation to addition—subtraction refers to the number of iterations of a unit (positive or negative) between one vector and the other.

Reflection

If <7> is seven iterations of the unit vector <1> and <−5> is five iterations of the unit vector <−1>, why does it follow that <7> + <−5> = <2> and <7> − <−5> = <12>?

Working with 1-by-1 vectors and 1-by-1 matrices, we come to see vector addition simply as addition and matrix multiplication simply as multiplication that affirms Brahmagupta's rules: a negative times a positive is negative, and a negative times a negative is positive. However, representing numbers as vectors can help make explicit the mental actions we could use to construct those numbers. For example, consider the number −5/3. We can construct this number from a unit of 1, represented as the unit vector <1>, which is the result of *unitizing*. Next, we might transform this unit into the unit fraction of 1/3, by *partitioning* it into three equal parts and *disembedding* one of them. We can represent this transformation with the matrix [1/3], and the matrix product, [1/3]<1>, yields the vector <1/3>. We can *iterate* that vector five times to produce <5/3>. Finally, matrix multiplication by the matrix [−1] *reflects* <5/3> over <0> to the vector <−5/3>.

From Chapter 2, we already know what will happen if we perform the same reflection twice, so we know that [−1][−1] = [1], or [−1]<−1> = <1>, or simply (−1)(−1) = 1. The first equation emphasizes 1 as the multiplicative identity, [1]; the second equation emphasizes 1 as a unit, <1>; both are important properties of the number, 1.

We need now contend only with multiplicative inverses (i.e., division) to fully integrate negative numbers into our newfound number system. Then, we will know how to add, subtract, multiply, and divide numbers, both positive and negative.

In *Mathematics: The Loss of Certainty*, math historian Morris Kline recounts an argument by a mathematician who questioned the meaning of division by -1. Considering the equality, $-1/1=1/-1$, he asked, "[H]ow could a smaller be to a greater as a greater is to a smaller?"[4] Quandaries like this arise when we generalize rules about mathematical objects rather than extending the actions that define them. For example, students rely on the rule that multiplying makes bigger, so they have trouble understanding fraction multiplication wherein the product is sometimes smaller than either fraction in the product (e.g., $1/2 \times 1/3 = 1/6$ is smaller than both $1/2$ and $1/3$). However, as seen in Chapter 4, we can resolve the issue when we consider fractions as coordinations of iterations and partitionings of units, and when we consider multiplication as a transformation of units.

We can address even the quandary of 1 divided by -1, by appealing to the mental action of reflection and the geometry of the number line. Division by -1 symbolizes the inverse action of multiplying by -1, which itself symbolizes a reflection and has itself as its inverse. Thus, $1/-1$ symbolizes the same object (that is, the same coordination of actions) that -1 does. We need not take numbers as given or rely on their formal properties; rather, we can appeal to the mental actions we used to construct those numbers in the first place.

COMPLEX NUMBERS

In the course of solving cubic equations with their newfound cubic formula, Italian mathematicians during the Renaissance generated expressions that contained square roots of negative numbers. Because these expressions, when combined, sometimes yielded real solutions, Italian Renaissance mathematicians like Cardano and Bombelli begrudgingly worked with them.[5] However, they weren't sure how to interpret the expressions themselves. Recall that, since ancient Egypt, Babylon, and Greece, $\sqrt{A}$ had referred to the side length of a square with area A. Now, with algebraic manipulation serving as a proxy for geometric construction, these Italian mathematicians had incidentally generated $\sqrt{-1}$. What could it mean; the side length of a square with area -1?

Outside of China and India, even negative numbers presented conceptual problems for mathematicians of yore. We just saw how to accommodate them, as directed quantities, by reflecting the positive side of the number line over the point 0. We can also accommodate imaginary numbers, like $\sqrt{-1}$, by moving in new directions.

When we included -1 into our system of numbers, we resolved its composition with other numbers, under addition and multiplication. Mathematicians initially tried to avoid the inclusion of negative numbers in algebraic equations by only considering polynomials with positive coefficients, but negative numbers appeared anyway and introduced this new problem of determining the meanings of their square roots. For example, even for the simple equation $x^2 + 1 = 0$ we encounter the positive and negative root of -1; symbolized as i and $-i$. If we want to accommodate these new numbers, once again, we have to ensure that they work with existing numbers in the system.

As a starting point, we know that $i^2 = -1$. We inherit this equation from the geometry of square roots—finding the side length of a square of area A. If the area is -1, and the side length is i (whatever that means), then the side length squared is -1.

We conceptualized multiplying by -1 as a reflection over the point 0, so now multiplying by i^2 is a proxy for that same mental action. Considering i^2 as i times i, how might we conceptualize multiplying by i? In other words, what mental action when composed with itself generates a reflection? There isn't such a mental action we can visualize in one dimension, but we can move into a second dimension, creating a plane of numbers.[6]

We can generalize the mental action of reflecting a one-dimensional number line about a point, 0, to rotating that number line π radians (180 degrees) about that same point, now the origin of a plane (see Figure 8.2) that we call the complex plane. Because the composition of two 90-degree rotations (π/2 radians) about the origin makes a 180-degree rotation about that same point, we can conceptualize multiplying a number by i as a 90-degree rotation of that number, about the origin.

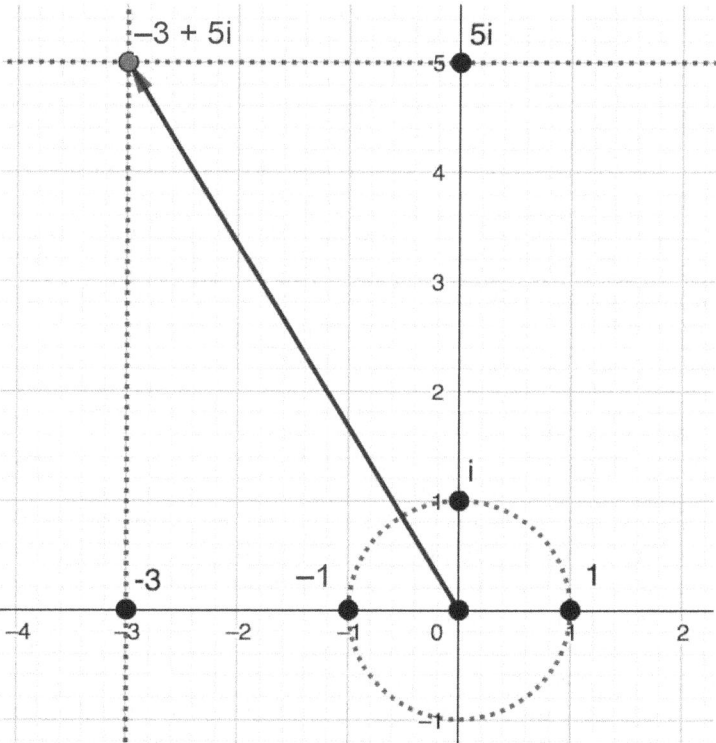

Figure 8.2 The complex plane.

Reflection

If multiplying a number by -1 yields a 180-degree rotation and multiplying by i yields a 90-degree rotation, what will multiplying the number by $-i$, i^4, and $\sqrt{i}$ yield?

With this 90-degree rotation introducing a new dimension of numbers, i becomes a new kind of unit, like 1. Now we can locate points in the plane as a (linear) combination of these units: 1 and i. We can iterate each of these units in either direction, forward or backward. As we saw with the (real) number line, iterations in opposite directions (1 and -1, or i and $-i$) undo each other, as additive inverses. Iterations in perpendicular directions simply adjoin the iterations, similar to joining 2 and 2 to make 4, but with iterations in their respective directions, contributing nothing to one another.

For example, Figure 8.2 illustrates the sum of -3 and 5i, where 3 is three units of -1 (or -3 units of 1) and 5i is five units of i. We can represent the result, $-3 + 5i$, as a vector showing the combined displacement from the origin to the final location in the complex plane. That is, we could represent $-3 + 5i$ as $<-3, 5>$, with real component -3 and imaginary component 5. In this way, we can locate any point in the complex plane as a vector with real and imaginary components. However, because the units for each component are contained within the same number system—the complex numbers—we can also think of $-3 + 5i$ as a single number.

From our iteration of units in different directions and compositions of those iterations, we generate a group of complex numbers, under the operation of addition. If we want to add a + bi and c + di, where a, b, c, and d are integers, we simply continue the iterations in each direction, $(a + c)1 + (b + d)i$. From Chapter 4, we know how to expand this number system to include fractional values, relying on the additional mental action of partitioning to create fractional units. At each stage, we are relying on the introduction of new mental actions to extend number, but then we must accommodate these new mental actions and the numbers they generate, within the closed system.

By now, you might have sensed that the introduction of new mental actions (e.g., introducing reflection to produce negative numbers on a number line) is not as challenging as the coordination of those new actions with other actions in the system. In the case of a complex plane, we know how to accommodate the 90-degree rotation and the imaginary numbers it generates, under the operation of addition. We simply add complex numbers component-wise: real units added to real units and imaginary units added to imaginary units. Now we have to contend with multiplying two complex numbers.

We have seen that vector addition preserves units in the sense that it continues the iteration of units in their respective directions. Multiplication transforms units so that these directions might change. In the product (a + bi)i, for example, the unit 1 is replaced with the unit i so that a + bi becomes a units of i plus b units of

i^2, or simply $-b + ai$ (it's the same result we would achieve by distributing i). More generally, $(a + bi)(c + di)$ replaces the unit of 1 with the unit of $c + di$, which is c units of 1 and d units of i. So, the product yields ac units of 1, ad units of i, bc more units of i, and bd units of -1, or $(ac - bd) + (ad + bc)i$. Notice how the units have been transformed, not just in terms of their magnitudes but also their directions. Figure 8.3 illustrates the geometric transformation.

We can represent this geometric transformation as a matrix, by noting how the two units, 1 and i, are transformed. We already know that 1 is transformed into $c + di$. Because 1 and i have a numerical relationship, we also know how i is transformed. Namely, it will go to $(c + di)i = -d + ci$; rotated 90 degrees from $c + di$. We can represent this transformation of units by the following matrix multiplication, multiplying the matrix by vector $<a, b>$ and resulting in the vector we might expect:

$$\begin{bmatrix} c & -d \\ d & c \end{bmatrix}\begin{bmatrix} a \\ b \end{bmatrix} = \begin{bmatrix} ac - bd \\ ad + bc \end{bmatrix}$$

We have generalized number in several ways, each time by introducing new mental actions to produce new units. In Chapter 4, we generalized whole numbers to fractions by introducing partitioning. Earlier in this chapter, we generalized whole numbers to integers by including a reflection. We can combine these two systems to form the rational numbers—a field of numbers we have extended to (rational) complex numbers by including i, along with a rotation of 90 degrees. We could extend further afield with new imaginary directions, creating the quaternions.[7]

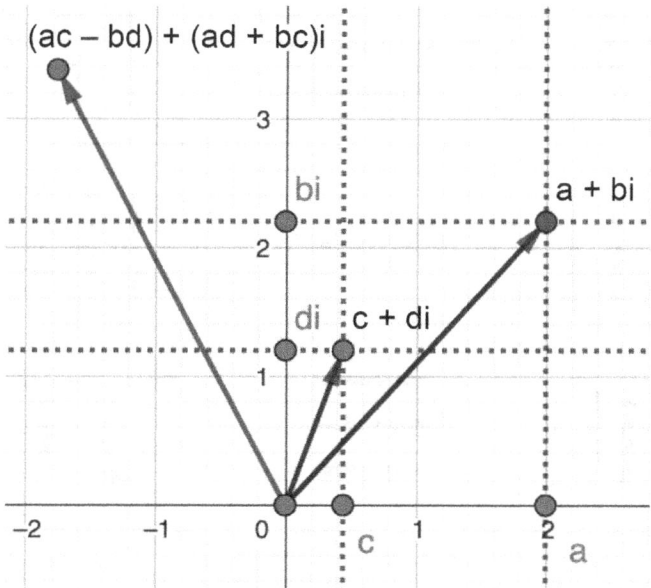

Figure 8.3 Complex multiplication as a unit transformation.

With each generalization, we have maintained two key formal operations: addition and multiplication. Addition preserves units, and multiplication transforms units. As we have begun to see, vector addition and matrix multiplication provide ways to extend those operations into higher dimensions.

TRANSFORMING THE SPACE OF NUMBERS

The Hindu-Arabic number system we use is a base-10 number system wherein each place value represents a power of 10, as a new level of units. For example, 321 represents one unit of 1, two units of 10, and three units of 100. Chinese mathematicians created the first known base-10 system, using an array to keep track of these various levels of units. They would place rods in columns where rods in the rightmost column represented units of 1, rods in the next column to the left represented units of 10, and so on. They would alternate the orientations of the rods—vertical and horizontal—to make the columns and powers of 10 easier to distinguish. Each row in the array could represent a different number. For example, the top two rows in Figure 8.4 represent the numbers 223 and 314. With their respective powers of 10 lined up, we can add the two numbers just by combining rods in a third row. Note that the π-shaped symbol represents 7, wherein the horizontal bar represents a grouping of 5 (used for numerals 6, 7, 8, and 9). If the sum in any column produced a grouping of 10, we would just carry it over as a

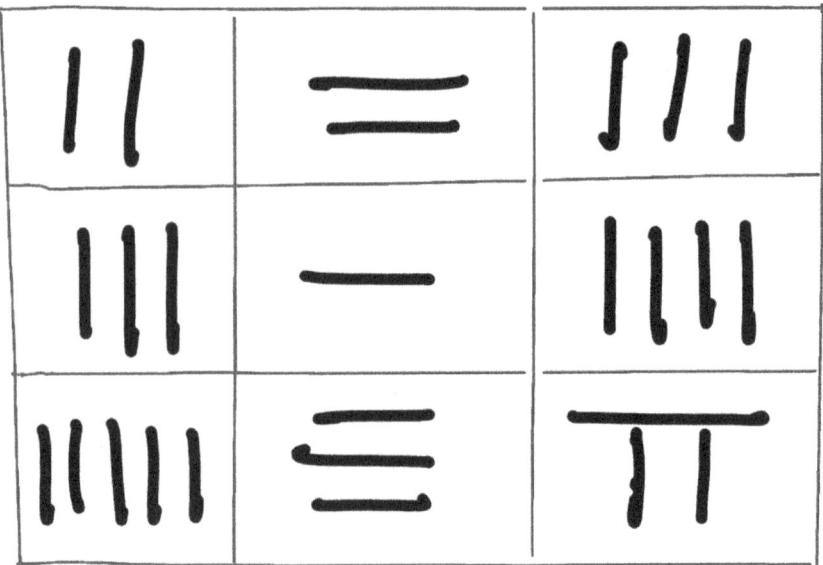

Figure 8.4 Addition using Chinese rods.

1 in the next column to the left, just as we do when we "carry the 1" using our Hindu–Arabic system.

By no mere coincidence, Chinese mathematicians also invented the method of row reduction for solving systems of equations.[8] Instead of representing the numbers 223 and 314, the top two rows in Figure 8.4 could just as readily represent the equations $2x + 2y = 3$ and $3x + y = 4$. Instead of units of 100, 10, and 1, the columns would represent (from left to right) units of x, y, and 1. By duplicating the numbers in each row and subtracting rows, Chinese mathematicians could eliminate unknown units until only one kind of unit remained. For example, by doubling the first row, tripling the second row, and then subtracting the second row from the first row, we find that $4y = 1$, so $y = 1/4$.

Today, we often use matrices and vectors to represent the equations as follows:

$$\begin{bmatrix} 2 & 2 \\ 3 & 1 \end{bmatrix} \begin{bmatrix} x \\ y \end{bmatrix} = \begin{bmatrix} 3 \\ 4 \end{bmatrix}$$

The unknown units, x and y, are represented within a vector and transformed by a 2-by-2 matrix (via matrix multiplication) into the resulting vector. So, we have a transformation of units, from x and y, to $2x + 2y$ and $3x + y$, which we know to equal 3 and 4, respectively. If we start from the vector <3, 4> and undo the transformation described by the matrix—multiplying the vector by the inverse of that matrix—we get values for x and y.

In extending numbers to higher dimensions, we focus on how matrices transform unit vectors, like <1, 0> and <0,1>. Those transformations determine a geometric transformation of the entire plane. In three-dimensions, the transformations of <1, 0, 0>, <0, 1, 0>, and <0, 0, 1>, using a 3-by-3 matrix, would determine how the matrix transforms all of space. We can use matrices to represent geometric transformations of four-dimensional space and higher, too, even if we can't visualize them. Here, we illustrate the idea in two dimensions.

Consider the following generic matrix, where a, b, c, and d are real numbers:

$$\begin{bmatrix} a & b \\ c & d \end{bmatrix} \begin{bmatrix} 1 \\ 0 \end{bmatrix} = \begin{bmatrix} a \\ c \end{bmatrix}$$

The matrix is multiplied by the unit vector <1, 0>, and matrix multiplication dictates the result: $<a \times 1 + b \times 0, c \times 1 + d \times 0> = <a, c>$. Likewise, multiplying the matrix by the unit vector <0, 1> yields $<a \times 0 + b \times 1, c \times 0 + d \times 1> = <b, d>$. We are often taught how to perform matrix multiplication in this way, as a rule to follow, but the rule itself follows something more essential. It is set up so that the first column tells us where the unit vector <1, 0> goes, and the second column tells us where the unit vector <0, 1> goes. In other words, matrix multiplication works in such a way that we can read the unit transformations right off the matrix itself. In turn, we can imagine the geometric transformation of the plane that those unit transformations induce.

Reflection

If $a = 0$, $b = 1$, $c = 1$, and $d = 0$ in the matrix shown earlier, where will the unit vectors $<1, 0>$ and $<0, 1>$ go? How would you describe the geometric transformation of the plane induced by matrix multiplication and this transformation of units?

When represented as vectors, every point in the plane, $<x, y>$, is just some combination of the unit vectors $<1, 0>$ and $<0, 1>$. In linear algebra, we call this combination of units a linear combination. Specifically, a generic vector $<x, y>$ is x iterations of $<1, 0>$ and y iterations of $<0, 1>$. So, if we know how a matrix transforms $<1,0>$ and $<0,1>$, we know how it transforms every point in the plane; it takes $<x, y>$ to $x<a, c> + y<b, d> = <ax + by, cx + dy>$. The transformation of the two unit vectors determines the transformation of the entire plane.[9]

In Chapter 2, we generated isometries of the plane, such as translations and rotations, from compositions of reflections. We can represent some of these isometries with matrices: reflections over lines passing through the origin, and rotations about the origin. We can use matrices to represent additional linear transformations, such as dilations from the origin (along either axis) and shears.[10]

Reflection

What would happen to the unit vectors $<1, 0>$ and $<0, 1>$ under a counterclockwise rotation of 90-degrees? Can you use this knowledge to find the matrix that would represent such a rotation?

Note that the unit transformation described in the reflection earlier also represents the effect of multiplying by i, as discussed in the prior section. Because the two units in the complex plane, 1 and i, have a numerical relationship, once we know how the unit of 1 is transformed, the transformation of i is also determined. Namely, if 1 goes to $<a, c>$, i goes to $<-c, a>$; the two vectors remain 90 degrees apart. Thus, as indicated by the following matrix, unit transformations induced by multiplying complex numbers are restricted to those represented by the following 2×2 matrix:

$$\begin{bmatrix} a & -c \\ c & a \end{bmatrix}$$

Geometrically, they include dilations of the entire plane from the origin, rotations about the origin, and compositions of those two transformations. These transformations are most readily described in polar coordinates: r (for the radius of the dilation) and θ (for the angle of rotation). In Chapter 10, we'll see connections between rectangular and polar coordinates, as represented by the equations $x = r\sin\theta$ and $y = r\cos\theta$. Here, we use rectangular coordinates.

TRANSFORMING AREA

In Chapter 5, in the context of proving the Pythagorean theorem, we investigated mental actions that create and preserve area. Specifically, independent sweeps create area, isometries (e.g., rotations), and shears preserve area. In transforming units, matrix multiplication can preserve, transform, or even destroy area. However, it cannot create area because it doesn't create new units—it only transforms existing units.

Figure 8.5 illustrates the transformation represented by the following generic matrix:

$$\begin{bmatrix} a & b \\ c & d \end{bmatrix}$$

In that transformation, <0, 0> goes to <0, 0>, <1,0> goes to <a, c>, <0, 1> goes to <b, d>, and <1, 1> goes to <a + b, c + d>. Note that the transformation of these four vectors show how the matrix transforms the unit square into a parallelogram.

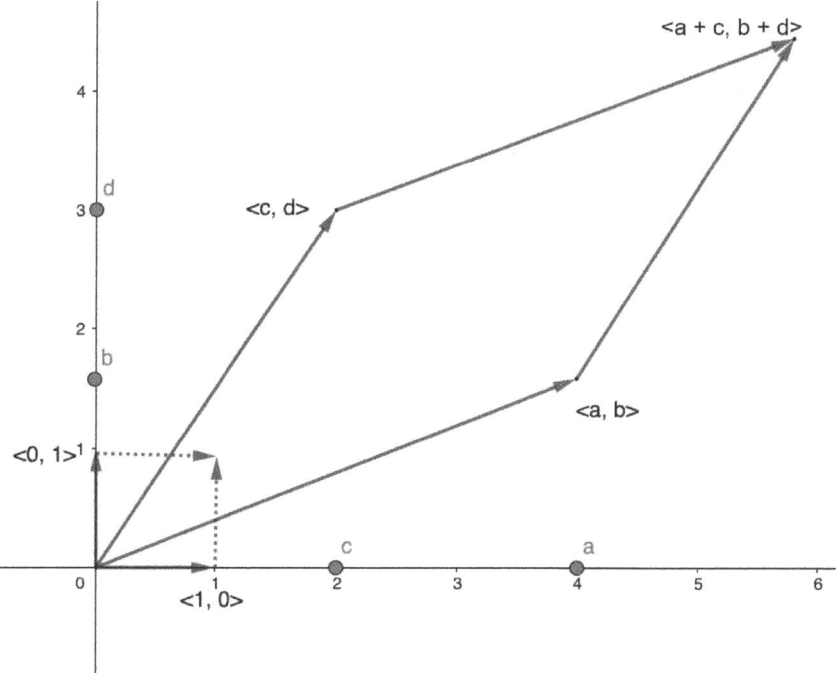

Figure 8.5 Matrix transformation.

The area of the unit square (1 square unit) is produced by the two independent (and perpendicular) sweeps indicated by the two unit vectors: <1, 0> and <0, 1>. When the matrix transforms these unit vectors, it also transforms the length and direction of their sweeps, which in turn determines the area of the parallelogram. What is its area? We could dissect the figure into triangular parts and use area formulas, but the goal here is to understand how area is transformed by mental actions, like shears.

Consider the two shaded rectangles in Figure 8.6. Note that if we sheared the right side of the larger rectangle up by c, or if we sheared the top side to the right by b, the area would be preserved. However, in a sense, we are making both transformations at once. The result is not a shear, or the composition of two shears, because as the two shears are performed they introduce new components to each side. The first shear, in the direction from (a, 0) to (a, c) would have been parallel to the left side of the rectangle, <0, d>, but it is not parallel to <b, d>, because the second shear, in the direction from (0, d) to (b, d) has introduced a component in the <1, 0> direction; and likewise for the effect of the second shear on the first shear.

We find ourselves in the same situation we were in with the generalization of the Pythagorean theorem from Chapter 5 (see Figure 8.7), with one key difference. In the generalized Pythagorean theorem, the horizontal shift of the top side (left, by length kb) restored perpendicularity of the two sweeps. In the present case, the horizontal shift of the top side (right, by length b), renders the two sweeps less independent.[11] Whereas the sides of the larger rectangle were perpendicular, the vectors <a, c> and <b, d> are more closely aligned. Specifically, the vector <a, c> has a component, c, in the vertical direction and the vector <b, d> contains a component, b, in the horizontal direction. The product of these components, b × c, is the amount of area lost by the reduction of independence. Thus, the area of the rectangle, a × d, is reduced by b × c when transformed into the parallelogram, which has an area of a × d − b × c, the determinant of the matrix.

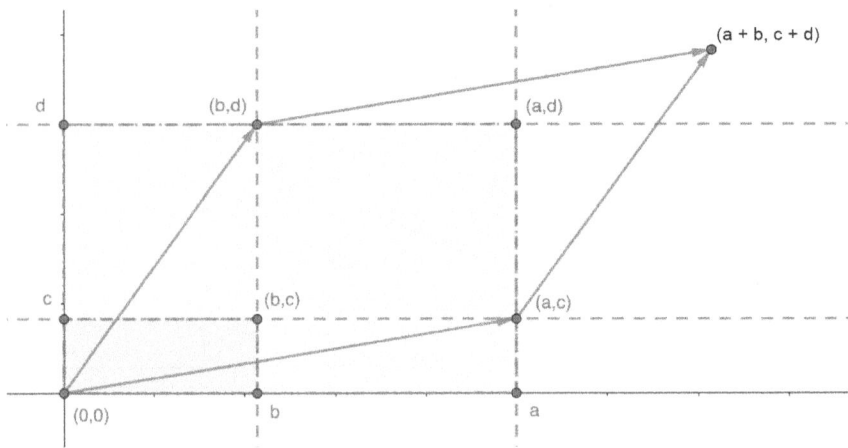

Figure 8.6 Determining area transformations with determinants.

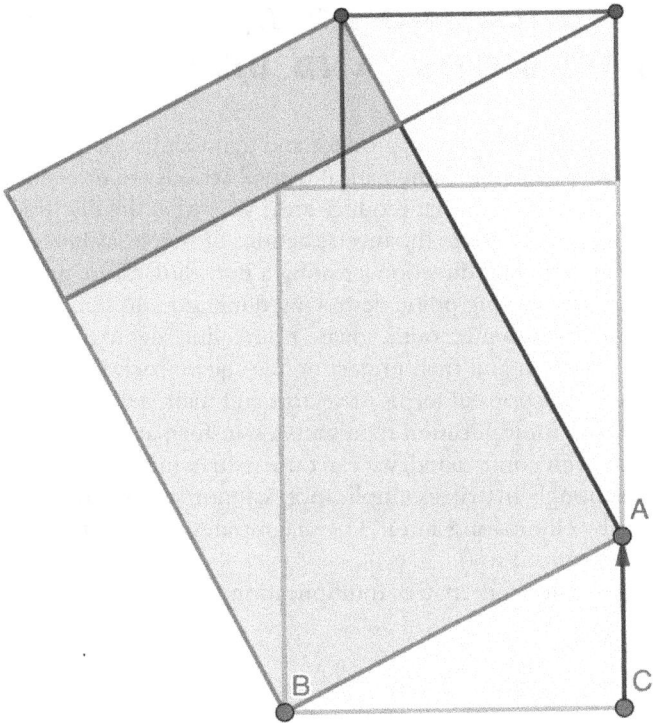

Figure 8.7 Generalization of the Pythagorean theorem.

By monitoring the transformation of units induced by matrix multiplication, we have demonstrated that the determinant of the matrix determines the transformation of area that the matrix multiplication induces. In the case of 2-by-2 matrices, the determinant is also the magnitude of the cross-product of the two transformed unit vectors, <a, c> and <b, d>, which, in general, measures the area between two vectors. We can understand it as a measure of perpendicularity between the two vectors.[12] More specifically, it is the product of the two vectors' magnitudes times the sine of the angle between them, and this value of sine measures the perpendicularity, orthogonality, or degree of independence of the two vectors. Thus, we find a formal connection between independent sweeps, constructing area, and the Pythagorean theorem (Chapter 5).

Formally, the Pythagorean theorem tells us that the areas of squares are "orthogonally additive": when a and b are orthogonal (perpendicular) vectors, the magnitude of a squared plus the magnitude of b squared equals the magnitude of a + b squared (note that, as a vector, c = a + b). To understand such connections in full, we would need to investigate them in more detail. To generalize them to higher dimensions, we would need to account for greater complexity. Here, we have established their psychological basis, as grounded in our own mental actions.

ANNIHILATING AREA, DESTROYING DIMENSIONS, AND DIVIDING BY 0

In considering mental actions that produce and transform area, we've left unfinished one critical discussion. If all mathematical mental actions are reversible, how are we to reverse the mental actions that produce area? We began the discussion in Chapter 5 by introducing projections as the inverse actions of sweeps. Whereas sweeps sweep out a length in a particular direction, creating a new dimension, projections project sweeps back to their starting point, destroying dimension. In Chapter 5, we considered the example of sweeping out a square from a line segment by moving along a perpendicular direction and then projecting the square back into the line segment. Here, we frame projections in terms of vectors and matrices.

Because matrix multiplication represents a transformation of units from a space that has already been constructed, we can't use matrix multiplication to create new dimensions. Although matrix multiplication cannot sweep out new areas, it can annihilate area by eliminating a unit. The magnitudes of all vectors in the direction of that unit are reduced to 0.

Consider the following matrix multiplication:

$$\begin{bmatrix} 1 & 2 \\ 1 & 2 \end{bmatrix} \begin{bmatrix} x \\ y \end{bmatrix} = \begin{bmatrix} 0 \\ 0 \end{bmatrix}$$

The matrix transforms units, from $<1,0>$ to $<1\ 1,>$, and from $<0,1>$ to $<2,2>$. The equation indicates that, as the two unit vectors are transformed, the vector $<x, y>$ is transformed to the zero vector, $<0, 0>$. The determinant of the matrix indicates that the area of the unit square is annihilated: $1 \times 2 - 1 \times 2 = 0$. How does this happen?

Figure 8.8 illustrates the transformation of units induced by the matrix multiplication. Note that, the two unit vectors get mapped into the same line. Because there is now only one unit direction, the entire plane is projected into that line. To identify the projection this transformation represents, we need only find the vector that gets mapped to the zero vector. In other words, we need to solve for the vector $<x, y>$ in the previous equation. In fact, $<x, y>$ represents the direction of a whole line of vectors that collapse to length 0.

Reflection

Find values of x and y that satisfy the equation above. Can you use pairs of values to determine the direction of the projections?

We refer to matrices with determinant 0 as non-invertible. If area is annihilated and dimension destroyed, there is no matrix that can re-create them. Although the matrix does not have an inverse, the mental action it represents does; namely,

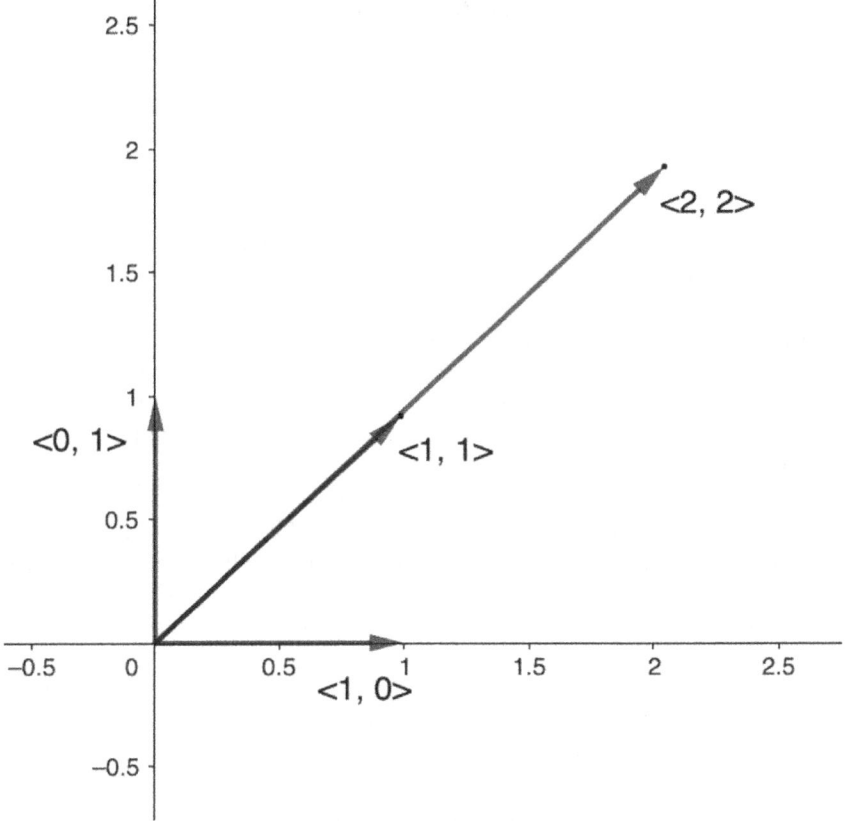

Figure 8.8 Destroying a dimension.

projections are reversed by sweeps. We see a simpler example in the context of multiplying integers.

Earlier in this chapter, we had to resolve the issue of multiplying and dividing by −1, in order to accommodate the expansion of whole numbers to integers. However, we never resolved the issue of multiplying and dividing by 0. The quandary we face is that multiplying by 0 entails transforming units of 1 to a unitless number. For example, 5 is five units of 1, but in the transformation 5 × 0, it becomes five units of 0. The product has no magnitude or direction and becomes an object of zero dimension.

We can think of multiplying by 0 as multiplying by the 1-by-1 zero matrix, [0], which has a determinant of 0. Thus, like all matrices that annihilate area and destroy dimension, multiplication by 0 represents a projection. In destroying the unit, multiplication by 0 projects all integers on the number line to the same point, 0. We can reverse the mental action of this projection, by sweeping out a line from that point. However, we can't uniquely determine the number on that line from which we began. So, although the mental action is reversible, the matrix is non-invertible, and division by 0 (the inverse of multiplying by 0) is indeterminant (i.e., undefined).[13]

FIELDS OF NUMBERS

Linear algebra elucidates the unique roles addition and multiplication play in operating on numbers. Addition continues the iteration of units and preserves units of each kind. When we add two vectors, we combine them component-wise, with respect to standard unit vectors, such as $<1, 0>$ and $<0, 1>$. In contrast, multiplication transforms units, replacing one unit with another. When we multiply a vector by a matrix, the vector generally changes magnitude and direction with respect to new units. It transforms along with the entire space.

We have seen that when we expand whole numbers to integers, we gain additive inverses. These numbers form a group under the operation of addition. In Chapter 4, we saw that expanding whole numbers to fractions introduced multiplicative inverses. These numbers form a group under the operation of multiplication. Combining the two systems (extending fractions in positive and negative directions) forms a structure known in abstract algebra as a field—a number system closed under both addition and multiplication.[14]

Additive and multiplicative worlds meet in the field of rational numbers, wherein every number has both an additive and multiplicative inverse, with one exception. Because 0 is unitless, we can't accommodate it with a multiplicative inverse. However, we can accommodate the inclusion of new units through field extensions. In fact, this is exactly what we did when we introduced i into the number system. In order to keep the number system closed, under addition and multiplication, we had to introduce more than a single number, i. We introduced all possible compositions of i, under additive and multiplicative operations, with itself and with the rational numbers. By insisting that the new system include all these compositions, we generated a field of complex numbers, a + bi, where a and b are rational numbers. In other words, we performed a field extension.

We can perform all kinds of field extensions in this way. For example, if we want to include $\sqrt{2}$ in our number system, we just combine it, additively and multiplicatively, in every possible way, with itself and every other number in the system, until we have generated the larger field containing a + b$\sqrt{2}$, where a and b are rational numbers. We return to this idea in Chapter 10, as we consider constructible polygons and constructible numbers.

SUMMARY

By introducing new mental actions of reflection and rotation, we have generalized whole numbers to integers and to complex numbers. The mental actions produce new units, in new directions (-1 and i). This idea is further generalized in linear algebra, wherein unit vectors reach higher and higher dimensions. Linear algebra elucidates the unity of space and number. The same mental actions used to construct geometric objects and transform space become mental actions for constructing new units and transforming them in space.

Introducing new actions and units also introduces new problems to solve. For example, we had to reconcile what it might mean to divide by −1 and multiply by i. In the end, there remained just one number we couldn't fully integrate into our number system—the one that has no units. Multiplying by 0 destroys not only area but units too. It represents a projection in which numbers in an entire dimension are collapsed onto one another. Reversing this action, through sweeping, would produce a one-to-many relationship between 0 and those numbers, which is why dividing by 0 is undefined. Nonetheless, the mental actions themselves—projecting and sweeping—are reversible.

Activities

Activity 1: Wawro, Rasmussen, Zandieh, Sweeney, and Larson (2012) designed an inquiry-oriented instructional activity called "the magic carpet ride," for college students to learn about linear combinations and span in linear algebra. The task involves two modes of transportation: a magic carpet that moves only in the direction (forward or reverse) described by the vector <3, 1> and a hoverboard that moves only in the direction described by the vector <1, 2>. Students are challenged to determine which locations on a map (the plane) they can reach using only these two modes of transportation. For example, how might they reach map coordinates (10, 0)? Are there any locations they can't reach?

Activity 2: One solution to the first question in Activity 1 is to travel forward on the magic carpet four times (4 × <3, 1>) and then to travel in reverse on the hoverboard twice (−2 × <1, 2>). If we think about the vectors <3, 1> and <1, 2> as units, we have four units of the first vector and negative two units of the second vector. However, <1, 0> and <0, 1> are the standard units we use to identify vectors in the plane (together we call them the standard basis for vectors in the plane). So, what happens when we transform those standard units into the new units, with the following matrix multiplication?

$$\begin{bmatrix} 3 & 1 \\ 1 & 2 \end{bmatrix}\begin{bmatrix} 4 \\ -2 \end{bmatrix}$$

What is the result and what does it represent?

Activity 3: Suppose you wanted to transform the plane by reflecting it over the x-axis and dilating it from the origin by a factor of 3. What matrix would represent this transformation?

$$\begin{bmatrix} a & b \\ c & d \end{bmatrix}$$

Recall that the determinant of a matrix, ad − bc, determines how the matrix transforms area. What is the determinant of the matrix you found and why does this value make sense?

Activity 4: Which matrix represents multiplying vectors in the complex plane by the complex number −3 + 5i? Use this matrix and matrix multiplication to determine the following products:

(−3 + 5i)(1 + 0i) (−3 + 5i) × (0 + i) (−3 + 5i) × (−3 + 5i)

How do these products make sense geometrically (see Figure 8.3)?

NOTES

1. Barrow-Green, Gray, and Wilson (2019) include a complete list of Brahmagupta's rules and an elaborate description of the early Indian number system on which our current Hindu-Arabic number system is based.

2. Mathematicians have variously referred to negative and imaginary numbers as "false," "fictitious," or "impossible" throughout history (Burton, 2007).

3. In their research on children's construction of numbers, Bofferding (2019) and Wessman-Enzinger (2019) each expand on this idea of integers as directed quantities.

4. Kline (1982, p. 11) shared this historical anecdote in his book, *Mathematics: The Loss of Certainty.*

5. In particular, Burton (2007) describes Bombelli's "wild thought," in the context of solving cubic equations, that led him to determine that complex roots come in conjugate pairs (a + bi and a − bi) so that their sums generate real numbers.

6. "[In his 1685 book, *Algebra*] Wallis said, in effect, that complex numbers are no more absurd than negative numbers and, since the latter can be represented on a directed line, it should be possible to represent complex numbers in a plane" (Kline, 1982, p. 118).

7. i is not the only direction we could have chosen in representing a square root for −1. All we needed was for the direction to be perpendicular to the real line, and in space, there are at least two such directions (not counting their negatives, such as −i). Conventionally, we represent these directions as the y- and z-axes within three-dimensional space. But why stop there? Although they might be difficult to imagine (see *Flatlands*, written by Abbott in 1899), there are limitless directions we could choose in higher dimensions. Quaternions are the result of choosing three imaginary directions, i, j, and k, to complement the real line.

8. See Andrews-Larson (2015) for an engaging review of the history of row reduction.

9. As long as <a, c> and <b, d> have different directions, this transformation is also a "change of basis." Whereas the standard basis for the real plane is given by the unit vectors <1, 0> and <0, 1>, linear combinations of <a, c> and <b, d> could also describe every vector in the plane

10. We can use 2-by-2 matrices (under matrix multiplication) to represent the group of isometries, excluding ones that involve translation. This group is isomorphic (structurally the same as) the group of invertible 2-by-2 matrices (2-by-2 matrices whose determinants are not 0). Note that translations are not included because the origin—represented by the vector <0, 0>—has no units and remains fixed under matrix multiplication.

11. As in Chapter 5, degree of independence refers to the relative amount by which one vector moves out of line with the first. In linear algebra, there is no formal distinction in the degrees of independence of vectors. However, we can interpret the sine of the angle between the vectors, as it appears in the cross product, as a measure of independence.

12. Likewise, we can understand the dot product as a measure of parallelism between two vectors.

13. If we allowed for division by 0, we would end up concluding that all numbers were equal to each other or lose the distributive property. See Norton (2016) for an elaboration on this discussion.

14. A field is like a group, except there are two operations to consider. With fields of numbers, we generally refer to the operations of addition and multiplication. Both operations are associative; there are additive and multiplicative identities (0 and 1); and every number in the field has both an additive and a multiplicative inverse, with the exception of a multiplicative inverse for 0. Fields also include the distributive property as a way of relating the two operations: $a \times (b + c) = a \times b + a \times c$.

REFERENCES

Abbott, E. A. (1899). *Flatland: A romance of many dimensions*. Boston: Little, Brown, & Company.

Andrews-Larson, C. (2015). Roots of linear algebra: An historical exploration of linear systems. *Primus, 25*(6), 507–528.

Barrow-Green, J., Gray, J., & Wilson, R. (2019). *The history of mathematics: A source-based approach: Volume 1*. Providence, RI: American Mathematical Society.

Bofferding, L. (2019). Understanding negative numbers. In A. Norton & M. W. Alibali (Eds.), *Constructing number: Merging perspectives from psychology and mathematics education* (pp. 251–277). Springer (Original work published in 2018).

Burton, D. M. (2007). *The history of mathematics: An introduction*. New York: McGraw-Hill.

Kline, M. (1982). *Mathematics: The loss of certainty*. New York: Oxford University Press.

Norton, A. (2016). *(Ir)reversibility in mathematics*. Proceedings of the Thirty-Eighth Annual Meeting of the North American Chapter of the International Group for the Psychology of Mathematics Education. Tucson, AZ: University of Arizona.

Wawro, M., Rasmussen, C., Zandieh, M., Sweeney, G. F., & Larson, C. (2012). An inquiry-oriented approach to span and linear independence: The case of the magic carpet ride sequence. *Primus, 22*(8), 577–599.

Wessman-Enzinger, N. M. (2019). Integers as directed quantities. In A. Norton & M. W. Alibali (Eds.), *Constructing number: Merging perspectives from psychology and mathematics education* (pp. 279–305). (2018). Springer (Original work published in 2018).

9

What Doesn't Vary as Variables Covary?

Emmy Noether grew up in Erlangen, Germany, at the turn of the 20th century. She graduated from, and then began working at, the university there—the same university where Felix Klein had begun his Erlangen program. As a consequence, she was drawn to mathematical research on algebraic invariants. As a woman, she worked without pay.[1]

Noether contributed greatly to the study of algebraic invariants, which is roughly the Erlangen program applied to algebraic objects (e.g., equations). The Erlangen program had classified geometries based on groups of transformations that leave their objects invariant. For example, although isometries and dilations move Euclidean objects (e.g., right triangles) to different positions in space, all internal relationships (angles and relative side lengths) remain the same. Thus, the group of isometries and dilations of the plane form the principal group for Euclidean geometry.

We discussed the Erlangen program in Chapter 3, along with a discussion of how we might extend the program to define geometric objects (e.g., quadrilaterals) based on groups of our own mental actions. In Chapter 4, we extended the idea further by defining fractions as a group of our own mental actions of partitioning and iterating (the splitting group). Then, in Chapter 6, we saw how we might use algebra to symbolize the actions and objects of arithmetic and geometry. Now, we consider how these algebraic expressions and equations themselves might become mathematical objects.

DOI: 10.4324/9781003181729-10

Like all objects, an algebraic object is something we can act upon, or transform in some way. Consider the quadratic equation $0 = ax^2 + bx + c$. If we substitute $-x$ for x (reflecting all values of x over 0, as described in Chapter 8), the equation will change, but its discriminant ($b^2 - 4ac$) will remain the same. Therefore, the discriminant of a quadratic equation is an algebraic invariant under this geometric transformation. This implies that the number and kind (real or imaginary) of roots of a quadratic equation are also unaffected by the transformation. In her dissertation, Noether identified hundreds of such invariants but with higher order polynomials and more complicated transformations.

Here, we do something similar. We will learn to treat algebraic equations as objects and investigate how they represent invariant relationships between the variables they relate. Even when we transform them through algebraic manipulation, the invariant relationship is maintained. Reasoning with equations in this way requires a particular kind of coordination of mental actions, called covariation. Marilyn Carlson first identified this kind of coordination and called it *covariational reasoning*: "the cognitive activities involved in coordinating two varying quantities while attending to the ways they change in relation to one another."[2] Covariational reasoning has far-reaching implications for understanding more advanced mathematical objects, such as derivatives, integrals, and differential equations.[3] It also enables us to interpret graphs, which represent the same invariant relationship between variables.

NEW SOLUTIONS TO ANCIENT PROBLEMS

The history of mathematics has advanced through problem-solving. We have already seen several examples of mathematical problems whose solutions prompted the invention of whole new branches of mathematics (after all, necessity is the mother of invention). For example, solving cubic equations led to the invention of imaginary numbers and complex analysis. We are about to see another example.

Reflection

Imagine lying in bed, watching a fly buzz around your room. How might you describe the path of the fly's flight; that is, how might you describe its position at any moment in time?

It would be impossible to describe the fly's path precisely to another person without some frame of reference. Your room itself might provide such a frame. Assuming it has corners and perpendicular walls, you could choose one corner of the room as the origin for the fly's path. Then, you might approximate the fly's distance from that corner at any time. The trouble with this approach is that there is an entire sphere's worth of positions in space that are the same distance from the corner, and one-eighth of them are in your room, so you have not described the fly's position precisely until you provide more information.

You might resolve the issue with spherical coordinates, describing angles up from the floor and over from one of the walls. You might also resolve it by using additional corners of the room. For example, if you knew the fly's distance from three different corners of the room, at any time, its position would be located as the intersection of three spheres. In fact, this is how GPS works, calculating your exact location on Earth from your distances from three or more satellites orbiting the Earth. Both solutions would work, along with many others, but here we consider a simpler solution.

Focus on the three walls that meet at a corner of the room. Rather than measuring distance from the corner, measure distances from the three walls. These distances refer to the shortest (perpendicular, or orthogonal) distances to each wall. You could use those three measures to pinpoint the fly's position at any time. In modern notation, we would describe those measures as $x(t)$, $y(t)$, and $z(t)$.[4] The three measurements vary in time, but they vary together, at the same time. For example, if the fly began its path from one of the walls and flew to the middle of the room 3 seconds later, we might have $x(0) = 0, y(0) = 2, z(0) = 3$ and $x(3) = 2, y(3) = 4, z(3) = 5$ (see Figure 9.1).

Legend says that French mathematician and philosopher Renee Descartes invented the cartesian plane, lying in his bed, just as we have imagined. History says he invented it while solving the Pappus problems. These problems

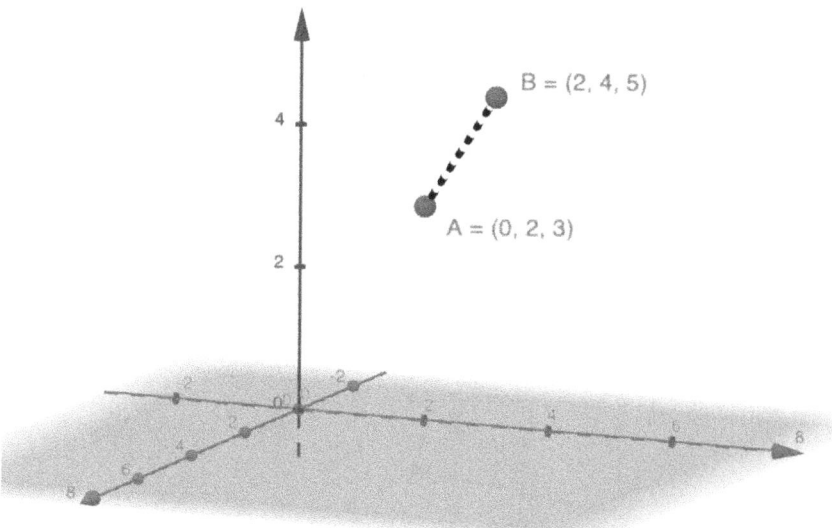

Figure 9.1 Locating flies.

date back to the 4th century when Pappus of Alexandria began posing them. They all involve finding a locus of points—a line, a curve, or a set of points— that satisfy given geometric conditions. Many of these problems remained unsolved until the invention of the cartesian plane. The following paraphrased (and temporally distanced) conversation, between Pappus and Descartes, tells the story:

> Pappus around 320 CE: *Given two parallel lines in a plane and a third line perpendicular to them, find the locus of a point that moves so that the product of the distances from the two first two lines is proportional to the square of the distance from the third line.*
>
> Descartes in 1631 CE: *I would simplify matters by considering the first line and the perpendicular line as the principle lines. Call the segment along the first line y, and call the segment along the perpendicular line x.*[5]

The problems Pappus posed presumed geometric constructions, but Descartes solved them by describing algebraic relationships between variables, x and y. Descartes had simplified matters by introducing these variables as distances along two of the given lines—variables that had the invariant relationship Pappus prescribed. Figure 9.2 illustrates Descartes' novel approach.

The first (leftmost) line and the perpendicular line serve as principal lines, or axes. They provide a frame of reference for describing all points in the plane. Specifically, we might call the first line the y-axis and the second line the x-axis. Distances from the y-axis could be measured as segments along the x-axis and vice versa; Descartes called these distances x and y, respectively. If we assume the two parallel (vertical) lines are one unit apart, then the product of distances from them would be $x(x - 1)$. The square of the distance from the third line (the x-axis) would simply be y^2. Satisfying Pappus's condition amounts to equating those two terms (possibly with the inclusion of a constant term), as graphed on the right side of Figure 9.2. This equation then describes the curve (locus) and the invariant relationship that Pappus prescribed.

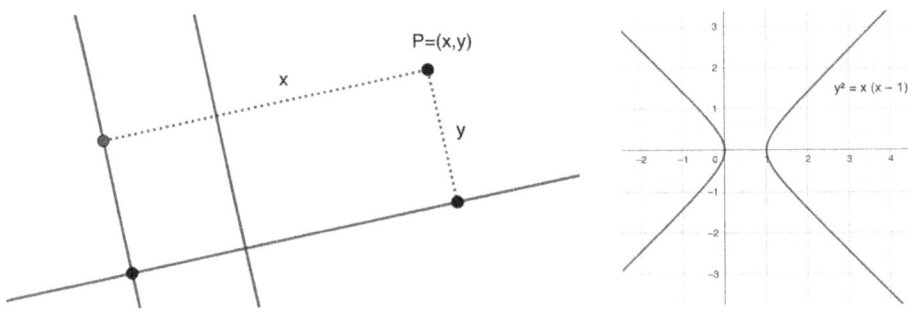

Figure 9.2 A Cartesian solution to an ancient problem.

Reflection

Consider the three "Pappus problems" that follow. How might you solve each of them using the Cartesian plane?

1. Given two perpendicular lines in a plane, describe all the points whose distance from one line is the same as its distance from the other line.
2. Given a point in the plane, find the locus of points 1 unit from the given point.
3. Given a line and a point not on the line, find the locus of all points equidistant from the given point and the given line

For the first problem, we might follow Descartes's lead in taking the two given lines as axes and measuring distances from each of them as x and y. Then the invariant relationship described in the problem is given by the equation $x = y$. For the second problem, we aren't given any lines, but if we choose perpendicular lines that intersect at the given point, and if we use those lines as axes, we can use the Pythagorean theorem to describe the locus. Every point in the locus would lie on a circle of radius 1, and the equation $x^2 + y^2 = 1$ describes the invariant relationship between the distances (x and y) to each axis. The third problem is more complicated and more closely aligned with the problems Pappus posed.

Ancient Greek mathematicians would solve this problem through geometric construction. Suppose we call the given point Q and the given line l. We could take an arbitrary point, R, on l and construct a perpendicular line passing through it (see Figure 9.3). Then, we could draw a line segment through Q and R and

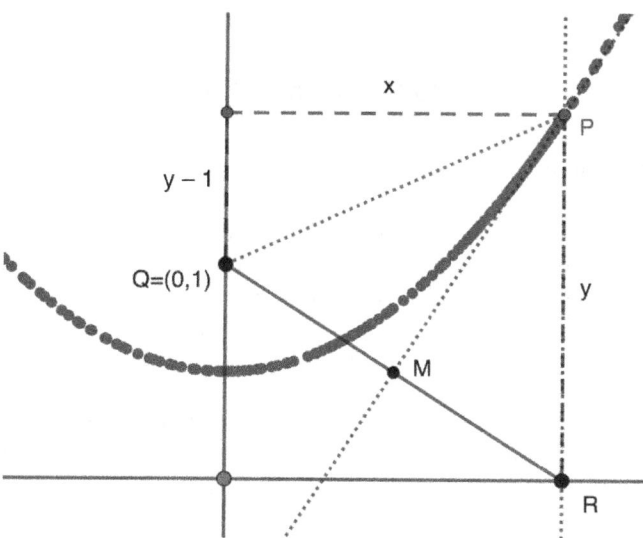

Figure 9.3 Constructing a parabola.

bisect it to find the midpoint, M. Finally, we could construct a line perpendicular to the line segment QR and passing though M. The intersection of the two perpendicular lines would be one of the points, P, on the locus. To find other points, we could just start the construction again from a different arbitrary point, R. We know these points satisfy the condition because the construction produces an isosceles triangle, QPR.[6]

In contrast to this geometric construction, Descartes's solution would rely on a pair of axes, relating the distances of points from those two axes within an algebraic equation. We could take the given line as the x-axis, and take the y-axis as the line perpendicular to it and passing through the point Q. If we take Q to be one unit from the x-axis, it would have the coordinates (0, 1). The coordinates for a point, P, on the locus would be given by (x, y)—its distances to the y-axis and x-axis, respectively. Now, we need to find a point, P, whose distance to 1 (the x-axis) is the same as its distance to Q: (0, 1). The former distance is just y; the latter distance is given by the Pythagorean theorem: $\sqrt{x^2 + (y-1)^2}$. So, the equation $y^2 = x^2 + (y-1)^2$. describes the curve. Algebraic manipulation of this equation won't affect the invariant relationship, so we could describe the same curve using the simpler equation, $y = \dfrac{x^2 - 1}{2}$.

The equation shown above and the graph shown in Figure 9.3 represent a parabola. Often students think about parabolas, as well as graphs of other curves, simply as shapes. However, as we have just seen, graphs and equations represent an invariant relationship between two co-varying quantities. If we want students to understand this dynamic relationship, rather than simply asking them to identify static shapes, we need to engage them in covariational reasoning.

COVARIATION

Understanding the cartesian plane involves much more than point plotting. If it were just a matter of plotting points on a grid, as if we were playing Battleship or Bingo, there would be little to celebrate. Descartes's idea was powerful because it described how two distances covary, allowing us to represent curves with algebraic equations. The power of this idea is available to anyone who can coordinate two varying quantities. As with all mathematics, this involves a coordination of our own mental actions. As with all quantitative reasoning, it also relies on the construction of units.

Imagine an hourglass draining sand (see Figure 9.4). What quantities vary in this situation, and how might we measure them? Let's say we want to focus on the height of the sand in the top part of the glass as one quantity, and the volume of sand in the bottom of the glass as another quantity. We might measure the height in centimeters (cm), and we might measure the volume in milliliters (mL). Then we might assign x(t) as the height in cm at a given moment in time, t, and we

Figure 9.4 Hourglass.
Source: © Eleanor Norton

might assign the variable y(t) as the volume in mL at that same given time, t. How might we describe the way these two variables, x(t) and y(t), covary?

Through their research on undergraduate students, Carlson and colleagues described a progression of five levels through which students learn to reason covariationally. Each level introduces a new mental action and coordinates it with mental actions at the lower levels. Table 9.1 summarizes these levels, mental actions, and their applications in reasoning with the hourglass task.[7]

In the situation described earlier, x(t) and y(t) both vary in time. However, we are interested in how they vary in relation to one another, not in relation to time. Taking time out of the equation, we can write the variables simply as x and y, so long as we keep in mind that each value of *x* corresponds with a particular value of y (the one that occurs at the same time). So, how do we represent the invariant relationship between values of x and y (at any given time)?

Suppose we wanted to represent this relationship with an equation. Reasoning at Level 1 wouldn't get us very far. At best, we might generate a table with a few values of x and y. Reasoning at Level 2 might lead us to guess an equation representing a direct and negative relationship, like y = −x; as x increases, *y* decreases. However, the situation does not become quantitative for us until we construct units and use them to measure the quantities as represented by lengths, or distances.

Reasoning quantitatively (Level 3), we might consider changes in rate. The rate of change is not invariant as x and y vary, so we need an equation that allows the

Table 9.1 Levels of covariational reasoning.

Level	*New Mental Actions*	*Reasoning With Hourglass Task*
Level 1: Correspondence	A simple correspondence between values of x and values of y at various times.	When the height is 10 cm, the volume will be 0 mL, and when the height is 0 cm, the volume will be 100 mL
Level 2: Direction	Sweeping out a length in a positive or negative direction to represent changes in x and noting corresponding directional changes in y.	As the height decreases, the volume increases.
Level 3: Quantitative Coordination	Partitioning the length representing values of x into equal units (say units of 1 cm each) and, for each unit, noting corresponding changes along the length representing y.	When the first cm of height is lost, the volume will increase by about 20 mL, but when the next cm is lost, the volume will increase less than 20 mL because the glass gets narrower.
Level 4: Rate of Change	Transforming the units representing equal changes in the values of x, into units representing changes in the values of y, establishing rates of change in values of y.	With each cm of height lost, the volume increases until it reaches 100 mL, but the rate by which it increases is less and less until it is 0.
Level 5: Instantaneous Rate of Change	Partitioning changes in the values of x into smaller and smaller units, and transforming them into changes in the values of y, thus refining the rates of change in values of y.	The rate of change in the volume is decreasing but the rate at which it decreases is constant because the glass gets narrower at a constant rate.

rate of change to change. To make an educated guess, we would need to focus on how the rate of change changes (Level 4). Specifically, the rate at which the rate of change changes is constant. This is the invariant relationship the equation should describe.

Notice that we are coordinating several levels of units here: units of x, measured in cm; units of y, measured in mL; also rates of change in y measured in mL/cm; and finally changes in rates of change of y, measured in mL/cm^2. At Level 5, we can take into account the smoothness of these changes and analyze their relationships with continuous functions—the subject of Chapter 11.[8]

Reflection

At what level do you find yourself? Can you improve upon the guess of y = −x, in describing the invariant relationship between height and volume?

We might build up an equation starting from an observation about changes in the rate of change. With every centimeter, the decrease in rate (measured in mL/cm) could be constant. In other words, the differences in the rate of change could be constant.[9] For example, suppose that when the height goes from 10 cm to 9 cm, the volume increases from 0 mL to 19 mL, and then when the height goes from 9 cm to 8 cm, the volume increases by 17 mL. The rate of change has decreased by 2 mL/cm. From that observation, we might infer something about the rate of change itself. In turn, the rate of change should tell us something about the equation. We return to this idea in Chapter 11.

Alternatively, we might represent the invariant relationship with a graph. Using perpendicular lines as a reference frame, we can create a Cartesian plane representing the two variables as distances from one line or the other. We can then trace a curve in the plane by coordinating these two distances. In coordinating these two distances, we have several cues to guide us, depending on the level of covariational reasoning at which we operation (see Table 9.1):

1. [Level 1] When the distance from the y-axis is 10 units (cm), the distance from the x-axis is 0 units (mL).
2. [Level 1] When the distance from the y-axis is 0 units (cm), the distance from the x-axis is 100 units (mL).
3. [Level 2] As the distance to the y-axis decreases, the distance from the x-axis increases.
4. [Level 3] For each of the first few units the distance from the y-axis decreases, the distance from the x-axis will increase relatively quickly.
5. [Level 3] For each of the last few units the distance from the y-axis decreases, the distance from the x-axis will increase slowly.
6. [Level 4] For each unit the distance to the y-axis decreases, the distance from the x-axis will rise more and more quickly.
7. [Level 5] The rate at which the rise occurs will change smoothly, at a constant rate.

Reflection

Use various levels of cues listed above to see how your graph evolves at higher levels of covariational reasoning. After you are done, compare your final graph to the one shown in Figure 9.5.

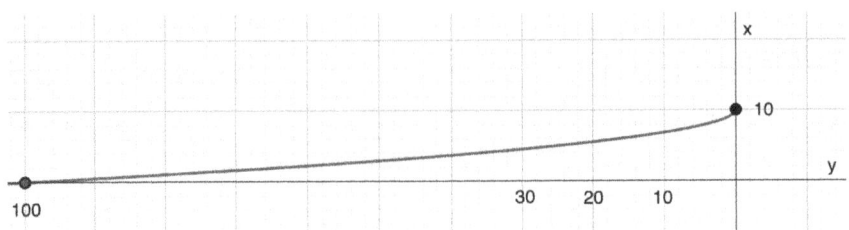

Figure 9.5 Cartesian plane.

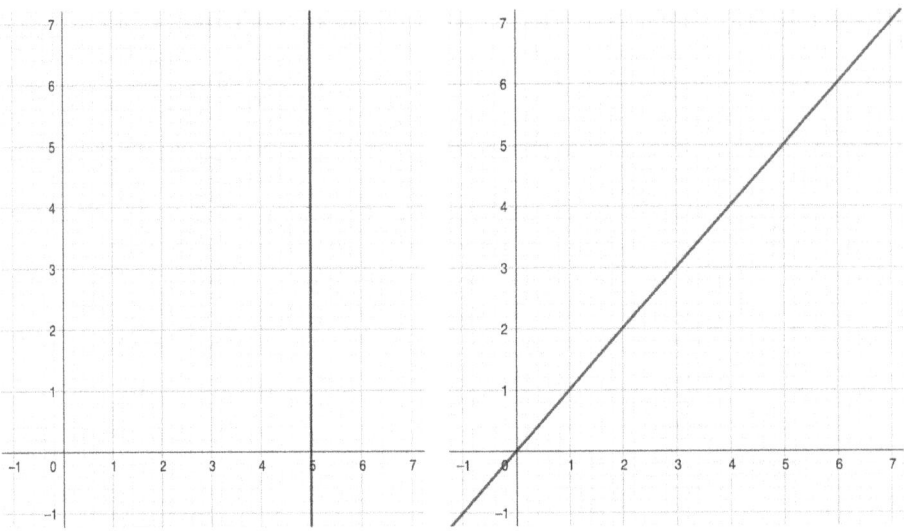

Figure 9.6 Rectangular coordinates.

The graph shown in Figure 9.5 is probably not the one you were expecting. In your experience drawing graphs, you have probably grown accustomed to certain conventions: the x-axis is supposed to be horizontal, and positive values of *x* are supposed to go to the right; the y-axis is supposed to be vertical, and positive values of *y* are supposed to go up. Mathematics is full of conventions, but unlike the coordinations of mental actions that produce mathematical objects, these conventions are not logically necessary. In fact, violating conventions can help us see what is logically necessary in a mathematical situation.

UNCONVENTIONAL GRAPHS
AND SHAPE THINKING

Let's re-interpret the graph shown in Figure 9.5. If you rotate it about the origin, clockwise by π/2 radians (90 degrees), the axes will be oriented as you might expect (positive values of x to the right and positive values of y at the top). However, the graph is just fine as it is—as it is shown in Figure 9.5—because it represents how the variables, x and y, covary. After all, representing covariation between variables was the entire purpose of the Descartes' plane.

Reflection

Consider the equations x = 5 and y = x. What graphs might represent the same covariation of variables?

Conventionally, we might say the graph of x = 5 should be a vertical line, through the point (5,0), where the *x* value is always 5 and the *y* value is free to vary, up and down. Likewise, we might say that y = x should be graphed as a diagonal line, where x- and y-coordinates always have the same value. However, we could interpret these equations differently by thinking differently about the variables they represent, focusing on how those variables covary and determining what in that relationship remains invariant.

What if x represented a distance from the origin (a radius) and y represented the measure of a central angle? Then the first graph would have a fixed radius of 5 while the angle would be free to vary. This relationship describes a circle (see the left side of Figure 9.7). Then the second equation would represent a covariation of measures of the radius and angle wherein the two measures remain the same (their invariant relationship). This would represent a spiral, in which the radius gets bigger and bigger as the angle goes around (see the right side of Figure 9.7).[10]

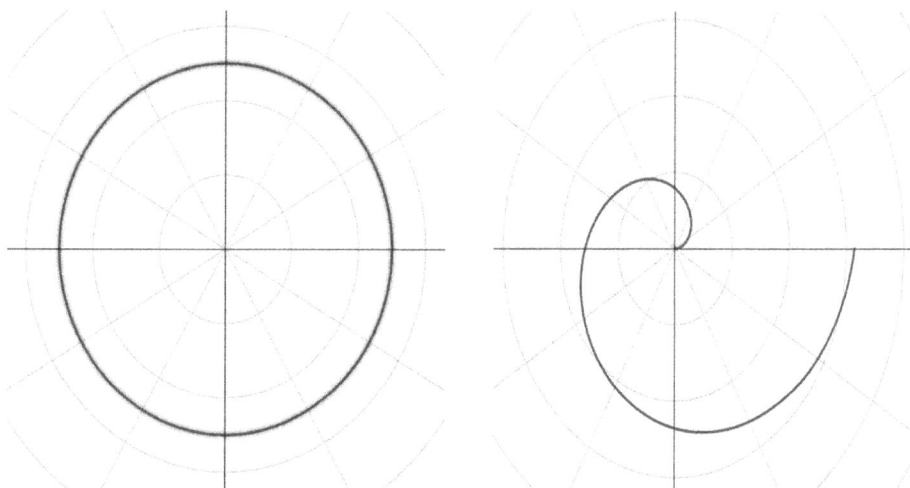

Figure 9.7 Polar coordinates.

A more conventional representation of these relationships, between angles and radii, would utilize polar coordinates—an alternative to the reference frame provided by the Cartesian plane. Rather than locating points in the plane by their distances from a pair of axes, polar coordinates locate them with a distance from a single point and an angle at that point (the origin). In the Cartesian plane, we rely on the Pythagorean theorem to describe a circle of radius 5 (similar to the unit circle, discussed in Chapter 10): $x^2 + y^2 = 5^2$. In polar coordinates, we could describe the same geometric object simply as $r = 5$. That equation represents the invariant relationship between the radius, r, and the angle, θ. Namely, the radius remains one unit long as the angle is free to vary. Likewise, in polar coordinates, we could describe the spiral as $r = \theta$.

When we look at graphs as if they were static figures, we cannot break free of convention. We might associate a parabola with a U-shaped graph, which might lead us to insist that (1) all U-shaped graphs are parabolas, and (2) if the graph is turned sideways, it can't be a parabola.[11] The challenge is to understand how graphs emerge through covariation. Then, a line might represent a circle or a spiral. Unconventional tasks like the following can support this kind of reasoning.

Suppose you wanted to take a cycling trip from Blacksburg, Virginia (home of the Hokies) to Washington, D.C. The shortest distance is 270 miles, but to avoid Charlottesville (home of the Wahoos), you decide to take a path like the one shown on the left side of Figure 9.8, which is 300 miles long. How might you describe your distance from Charlottesville during each mile of our trip?[12]

Reflection

Imagine or draw a graph representing the two co-varying quantities, with the x-axis representing the distance from Charlottesville and the y-axis representing total distance traveled.

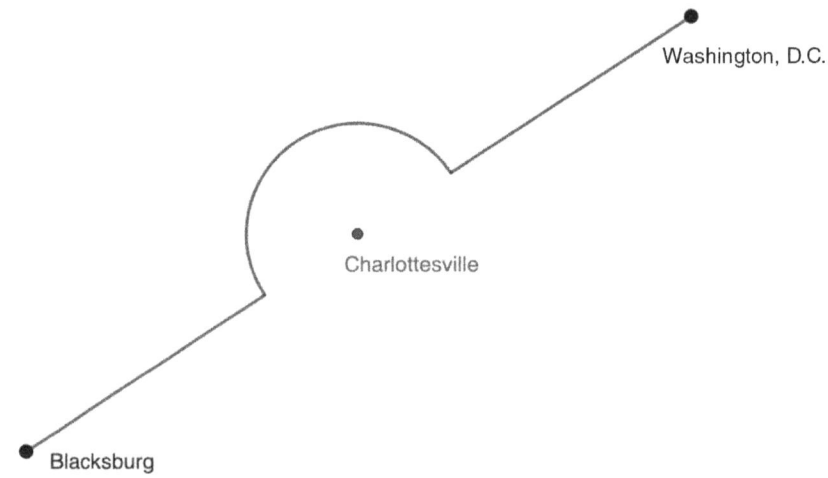

Figure 9.8 Driving to Washington, D.C.

Once again, you can rely on covariational reasoning to guide you. In this case, you are considering the distance from Charlottesville as a function of the total distance traveled. This means that each distance along your trip corresponds with a single distance from Charlottesville. When you have traveled 0 miles, you are about 135 miles from Charlottesville, and when you have traveled 300 miles, you are again about 135 miles from Charlottesville. How do those two distances covary in between? How do you represent the relationship between those distances as we travel along the bypass (semicircle)?

In graphing the distance from Charlottesville as a function of the total distance traveled, you will violate another convention: the vertical line test. You might have learned in school that you can determine whether a graph represents a function by drawing vertical lines through the graph. If any vertical line crosses the graph in more than one place, the graph is not the graph of a function, or so the story goes. The test works under certain conventions for graphing.

In reasoning covariationally, it's useful to let one variable (the independent variable) vary freely and consider how values of the other variable (the dependent variable) depend on it. We can see this kind of reasoning in Carlson's framework (Table 9.1). We typically take distances from the y-axis (distances along the x-axis) as representing the independent variable. Then we take distances from the x-axis (distances along the y-axis) as representing the dependent variable. This convention is what we mean by $f(x) = y$: that x represents the independent variable and that y will be a function of x (i.e., y will depend on x). Conventionally, we also orient the y-axis vertically and the x-axis horizontally.

Under these conventions, if a line passes through two points on the graph, there must be two distances along the y-axis associated with a single distance along the x-axis. That suggests that a single value of the independent variable gives two different values for the dependent variable. However, functions are supposed to have a single value for each value of the independent variable. Thus, conventionally, if a graph fails the vertical line test, it cannot be the graph of a function.

In the present case, the distance from Charlottesville is supposed to be a function of the total distance traveled, and in fact, it is. For every value of the distance traveled, there is a single value for the distance from Charlottesville. However, the graph violates the aforementioned conventions, so it fails the vertical line test, or rather, the vertical line test has failed. Especially when we consider graphs in polar coordinates or higher dimensions, the vertical line test becomes another rule to toss away. After all, mathematics is not a set of rules to follow. It is a product of your own reasoning (e.g., covariational reasoning) that explains rules and persists when those rules fail.

SUMMARY

In human experience, time is the ever-present variable in which we find ourselves. Like space, we construct our concept of time so early in life that we take it for granted. It sweeps through our lives, just as a point sweeps into a line. It's no wonder, then, that we take time as our fundamental intuition for variation.[13]

We sometimes refer to a variable as continuous, when what we mean is that the variable is swept out like time, as if it were swept out *in* time. In the example of a fly's path, we referred to three such variables: x(t), y(t), and z(t). Each of those variables is actually a function—a function of time: every moment in time corresponds with a value of x, y, or z. That is why we can think about those variables as continuous; continuity is a property of functions, not variables. We describe the function x(t) as continuous if changes in x are swept out like time itself, albeit with the possibility of changing direction or changing rate, with respect to time. For example, we might relate the passage of time to our position along a path. In that case, we would have a continuous function, x(t), relating position, x, and time, t.

When we consider the covariation of two variables, we can take time out of the equation. We understand that values of x and y correspond to one another, as if those values occurred at the same time. For example, in the hourglass problem x = 10 corresponds with y = 100, and x = 0 corresponds with y = 0. We don't need to explicitly refer to time to describe their relationship. However, we do need some way of measuring each of the two quantities through the iteration of units.

Although lengths we might sweep out a length continuously, we measure those lengths in discrete chunks, additively, by iterating a unit. For example, we can measure time by counting heartbeats, but each iteration of that unit is interrupted by a pause (just as von Glasersfeld described in his attentional model of units, discussed in Chapter 1). No matter how small our unit of measure, we still end up measuring continuous quantity through the iteration of some discrete unit.

The ancient Greeks grappled with this tension between continuous quantities and discrete units of measure. Like time, they took the continuum (what we now call a number line) for granted, as a line swept out in space. In their geometric constructions, they began with a unit of length 1, with which to create all other units and measure all lengths (or distances) along the continuum. They assumed they could construct all such lengths and had good reason for believing so. After all, they could construct all positive rational numbers, which densely fill the continuum, approximating every possible length as closely as they would like. Because decimals are fractions, we do this every time we write a decimal approximation for a number (e.g., $\sqrt{2} \approx 1.41$, also known as 141/100). However, we now know the Greeks could never complete such a program. Ultimately, their goal was replaced by Hilbert's formalist program, which relies on limiting processes, such as Dedekind cuts, to define the continuum as an uncountable collection of points.[14]

Notice that this history follows a progression similar to the levels of covariational reasoning Carlson described, wherein students partition a continuous quantity into finer and finer units for measuring it so that they can approximate an instantaneous rate of change. Many of the mental actions involved are familiar from previous chapters and earlier mathematical constructions: sweeping, unitizing, iterating, and partitioning. We return to this coordination of mental actions within a deeper investigation of continuity and rate of change, in Chapter 11.

The main goal of this chapter has been to understand equations and graphs, not as static forms, but as representative of dynamic relationships. Equations represent an invariant relationship between two variables as they covary. The cartesian plane and other coordinate systems (e.g., polar coordinates) provide us with a way to visually represent that invariance, as well as the variation. Graphs are not static shapes to describe as if they were mere pictures. They represent the way that two variables change in concert with one another. As we saw with the fly's path, we can extend this idea to three or more variables. Each time, the meaning of the representation—whether an equation or a graph—is provided by the units we construct to measure quantities and the mental actions we use to coordinate those measurements.

Activities

Activity 1: Invent your own coordinate system for locating points in the plane, using two variables and some kind of reference frame (e.g., distance from a fixed point, a fixed line, or a fixed circle)? How might you use that reference frame to describe a linear path, or a circular path in the plane?

Activity 2: Solve the following Pappus-styled problem by providing an equation for the locus: Given two perpendicular lines in a plane, describe all the points whose product of distances from the two lines is 1.

Activity 3: What invariant relationships do the following equations describe? Can you graph those relationships?
$x + y = 0 \quad x - y = 0 \quad xy = 1 \quad x/y = 1$

Activity 4: Returning to the hourglass problem, how would the situation change if we focused on the quantity of time, such as the way the volume changed as time changed, or the way the height changed as time changed? Could you approximate these relationships with an equation or a graph?

NOTES

1. Agnes Scott University provides a nice summary of Noether's life and work: www.agnesscott.edu/lriddle/women/noether.htm. For a thoroughly insightful biography, see Rowe (2021).

2. This quote comes from Carlson, Jacobs, Coe, Larsen, and Hsu's (2002, p. 354) article reporting on research on the covariational reasoning of college calculus students. That article also introduces the theoretical framework for covariational reasoning used in this chapter.

3. For example, see Rasmussen's (2000) research on college students' reliance on covariational reasoning for understanding differential equations.

4. These are called parametric equations. Although this notation was not available to Descartes, it fits his conception of each distance changing with time.

5. This fictionalized conversation is adapted from the actual language of Pappus and Descartes, as shared by Burton (2007).

6. As you argued in Activity 1 of Chapter 3, the perpendicular bisector of the base of an isosceles triangle passes through the third vertex of the triangle.

7. Table 9.1 is adapted from the covariation framework (Carlson, Jacobs, Coe, Larsen, & Hsu, 2002).

8. Castillo-Garsow, Johnson, and Moore (2013) make a distinction in students' reasoning, between chunky change, which changes in discrete units, and smooth change, which changes continuously.

9. Amy Ellis (2011) reports on a study in which middle school students invent the term *DiRoG* to describe differences in rates of growth that they notice while investigating quadratic equations.

10. Moore, Paoletti, and Musgrave (2013) used similar tasks to study the covariational reasoning of pre-service mathematics teachers.

11. Not all U-shaped graphs are parabolas. $y = x^4$ also has a U-shape, as does the catenary, which is a curve that describes how a rope, or chain, hangs between two points. Conversely, there is no reason they need to be oriented up (or down) like a U, especially when we consider the Greek construction of parabolas from a given point and a given line, which could be oriented in any way whatsoever.

12. Carlson's research on covariational reasoning, along with Pat Thompson's research on students' quantitative reasoning, has launched several new lines of research about how students make sense of graphs, especially in unconventional or unfamiliar settings. This task comes from researchers in that lineage: Moore, Stevens, Paoletti, Hobson, and Liang (2019).

13. Descartes, Kant, and other famed thinkers took time as innate—a concept with which we are born. Although we might take the passage of time for granted, measuring time is another matter. It depends on some other process, such as the ticking of a clock, the vibrations of a cesium atom, or the beating of your heart. Just think how inaccurate your estimations of time would be if you could not relate them to the passage of observable events.

14. Mancosu (1998) captures the historical debate between formalists like Hilbert, and intuitionists, like Brouwer, in his book, *From Brouwer to Hilbert*.

REFERENCES

Burton, D. M. (2007). *The history of mathematics: An introduction*. New York: McGraw-Hill.

Carlson, M., Jacobs, S., Coe, E., Larsen, S., & Hsu, E. (2002). Applying covariational reasoning while modeling dynamic events: A framework and a study. *Journal for Research in Mathematics Education, 33*(5), 352–378.

Castillo-Garsow, C. W., Johnson, H. L., & Moore, K. C. (2013). Chunky and smooth images of change. *For the Learning of Mathematics, 33*(3), 31–37.

Ellis, A. B. (2011). Generalizing-promoting actions: How classroom collaborations can support students' mathematical generalizations. *Journal for Research in Mathematics Education, 42*(4), 308–345.

Mancosu, P. (1998). *From Brouwer to Hilbert: The debate on the foundations of mathematics in the 1920s*. New York: Oxford University Press.

Moore, K. C., Paoletti, T., & Musgrave, S. (2013). Covariational reasoning and invariance among coordinate systems. *The Journal of Mathematical Behavior, 32*(3).

Moore, K. C., Stevens, I. E., Paoletti, T., Hobson, N. L., & Liang, B. (2019). Pre-service teachers' figurative and operative graphing actions. *Journal of Mathematical Behavior, 56.*

Rasmussen, C. (2000). New directions in differential equations: A framework for interpreting students' understandings and difficulties. *Journal of Mathematical Behavior, 20,* 55–87.

Rowe, D. E. (2021). *Emmy Noether—mathematician extraordinaire.* Cham, Switzerland: Springer.

10

What's So Special About Special Triangles?

High school trigonometry textbooks often refer to two special triangles: the 30–60–90 triangle and the 45–45–90 triangle. What makes these triangles special?[1] The short answer is that they allow us to geometrically determine the coordinates of points on the unit circle for a given angle; what we might call "nice coordinates" on the unit circle. Recall that the unit circle is simply a circle with a radius of 1 unit (hence the name *unit* circle), centered at the origin of the plane, (0, 0). When these two special triangles have their 30- and 45-degree angles at the origin, with one of their sides along the positive x-axis and with the hypotenuse as a radius of the circle, we can readily determine the x- and y-coordinates where that radius terminates along the unit circle. For example, consider triangle AOA' shown in Figure 10.1.

Angle AOA's measure is 60-degrees. Triangle sides OA and OA' are congruent—as radii of the unit circle—so their base angles, OAA' and OA'A, are also congruent. Thus, we have an equilateral triangle, with all sides of length 1. This triangle (triangle AOA') has the x-axis as a line of symmetry so that the top half of it, triangle AOX, is a 30–60–90 triangle with a hypotenuse of length 1 and side AX of length 1/2. Using the Pythagorean theorem (see Chapter 5), we can determine the remaining side length, OX, to have length $\sqrt{3}/2$. Therefore, point A has coordinates $(\sqrt{3}/2, 1/2)$. For the 45–45–90 triangle, and to find the coordinates of point B, we can do something akin to this with the square shown in Figure 10.1.

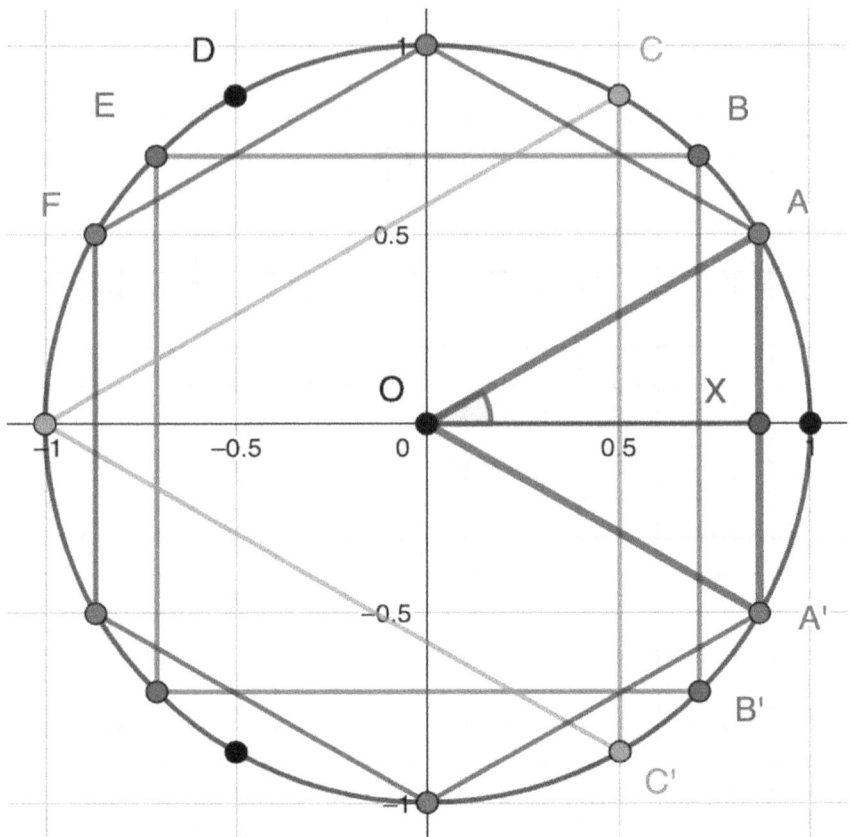

Figure 10.1 Special triangles.

Given all that we have considered about angle measure in Chapter 7, it might be more appropriate to refer to these special triangles as π/6–π/3–π/2 and π/4–π/4–π/2 triangles. This way, we can measure the central angles in the unit circle with lengths (arc lengths), just as we measure x- and y-coordinates on the unit circle with lengths. In fact, as shown in Figure 10.1 and recorded in Table 10.1, we can use the two special triangles to derive several lengths around the unit circle.

The lengths listed in Table 10.1 ground trigonometry in measurement. High school students typically learn trigonometric functions, such as sine and cosine,

Reflection

Use special triangles, the Pythagorean theorem, and symmetry, to determine the coordinates of points B, C, D, E, and F, in Figure 10.1.

as ratios of sides in a right triangle (e.g., "Sine is Opposite over Hypotenuse," as in SOHCAHTOA). This works as a rule but obscures trigonometric functions as *functions* that define how two quantities covary (see Chapter 9). In contrast, beginning with the lengths constructed from the two special triangles (Table 10.1), we can understand how trig functions emerge from a coordination of our own measurement activity.[2] For example, we might ask how the central angle and y-coordinate (height) on the unit circle covary, and Table 10.1 provides us with several benchmarks, represented by black dots in Figure 10.2.

Reflection

Do the black dots shown in Figure 10.2 make sense as benchmarks for height as a function of central angle? In other words, do they provide markers for how the central angle and y-coordinate covary on the unit circle? You might trace your finger along the unit circle and imagine the continuous curve these two covarying quantities form.

Table 10.1 Central angles and coordinates on the unit circle.

Central Angle	x-Coordinate	y-Coordinate
0 radians	1	0
π/6	√3/2	1/2
π/4	√2/2	√2/2
π/3	1/2	√3/2
π/2	0	1
2π/3	−1/2	√3/2
3π/4	−√2/2	√2/2
5π/6	−√3/2	1/2
π radians	−1	0
7π/6	−√3/2	−1/2
5π/4	−√2/2	−√2/2
4π/3	−1/2	−√3/2
3π/2	0	−1
5π3	1/2	−√3/2
7π/4	√2/2	−√2/2
11π/6	√3/2	−1/2
2π radians	1	0

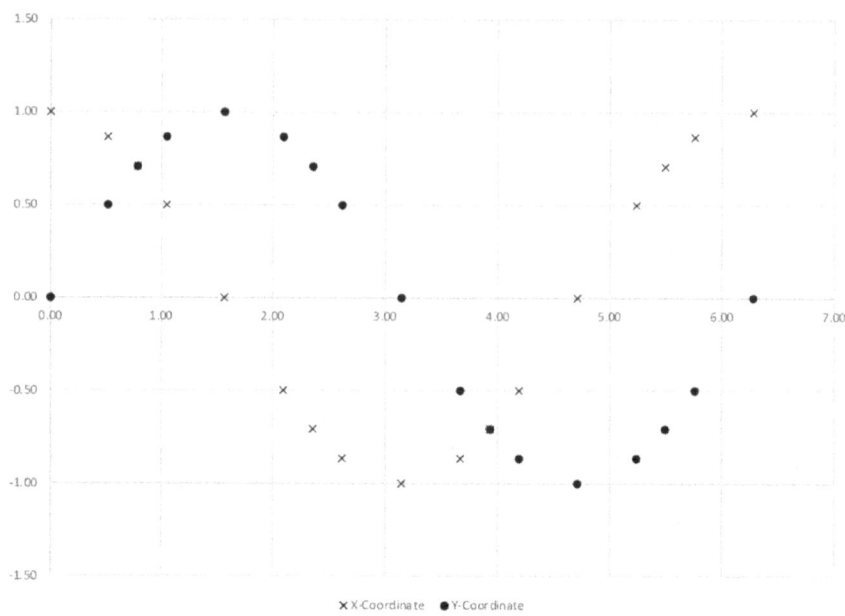

Figure 10.2 x- and y-coordinates, versus central angle, on the unit circle.

Your activity in coordinating these two covarying quantities generates the sine function. You might say that the sine function existed long before you engaged in this activity, but where? We can draw sine curves, write an equation for the sine function, or point to physical phenomenon modeled by it (e.g., the length of days over the course of a year), but like all other mathematical objects, the function does not live there.

> Socrates: And do you not know also that although [students of geometry] make use of the visual forms and reason about them, they are not thinking of these, but of the ideals which they resemble; not of the figures which they draw, but of the absolute square and the absolute diameter, as so on—the forms which they draw or make . . . they are really seeking to behold the things themselves, which can only be seen with the eye of the mind?
>
> (Plato's *Republic* Book VI, 510c[3])

Where do mathematical objects reside, if not in the figures we draw and the equations we write? Do mathematical objects have their perfect existence in the heavens, casting their shadows on the earth so that we, as mere mortals, might catch a glimpse? Alternatively, we could say that mathematical objects, such as the sine function, have no home except in the minds, brains, bodies, and societies of those who engage in and reflect upon the mental actions that define them. Like prior chapters, this chapter offers invitations to take up such activity, setting aside further philosophical inquiry until Chapter 12.

WHAT ABOUT COSINE, TANGENT, AND THE OTHERS?

The sine function comes into being through reflection on our own activity, coordinating heights and angles on the unit circle (as suggested in the earlier reflection). Once we have constructed the sine function, how can we come to know the other trigonometric functions: cosine, tangent, cotangent, secant, and cosecant? As you might imagine, for cosine, we can engage in the same activity as before, except with x-coordinates instead of y-coordinates. The x's in Figure 10.1 indicate associated benchmarks for the two covarying quantities: radian measures of the central angle and horizontal distance from the origin on the unit circle. All other trig functions are defined by, and derived from, sine and cosine. In fact, as seen in Figure 10.1, cosine is the same function as sine, shifted to the left by an angle measure of $\pi/2$; that is, $\cos(x) = \sin(x + \pi/2)$. So, we could define all trig functions in terms of sine.

You might know the tangent function as the ratio of the sine and cosine functions.[4] How might we interpret this ratio geometrically? Given what we know about sine and cosine, we can say it is the ratio of the y- and x-coordinates on the unit circle. In other words, it is the y-coordinate measured in units of the x-coordinate. This should also make clear that it is the slope of the radius of the unit circle for a given central angle. We investigate this idea further later.

Reflection

If we understand it as described above (i.e., slope of radius), what benchmark values can you derive for the tangent function?

Figure 10.3 focuses attention on the tangent function at $\pi/4$. The corresponding point on the unit circle is $(\sqrt{2}/2, \sqrt{2}/2)$, which in decimal form is approximately $(0.707, 0.707)$. At that point, it is clear that the tangent function should have a value of 1, but how can we see this in the figure, and why do we call the tangent function the *tangent* function?

As noted, the tangent function gives the slope of the radius for a given central angle. In this case, the radius extends from the origin, $(0, 0)$, to the point $(0.707, 0.707)$. The dashed line in Figure 10.3 is the tangent line to the circle at the point $(0.707, 0.707)$. You might notice it is perpendicular to the radius. As we will see in Chapter 11, this is always the case and implies that the slope of the tangent line is the negative reciprocal of the slope of the radius. In other words, the slope of the tangent line for a given angle is -1 over the value of the tangent function for that angle. Figure 10.3 also represents the tangent function as a length, from the point $(0.707, 0.707)$ on the circle to the point $(1.414, 0)$ on the x-axis. Can you show that this length is 1?

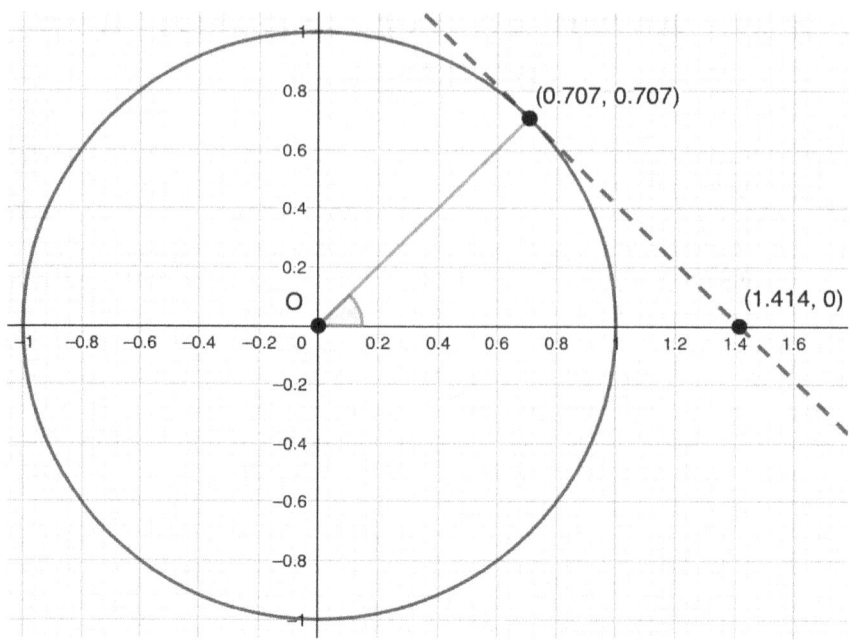

Figure 10.3 Tangent function and tangent line.

We could also represent secant and cosecant as lengths,[5] but here we rely on sine and cosine to define all other trig functions and to determine their values. In the definitions that follow, A represents the measure of the central angle (in radians, of course):

Tangent: ratio of sine to cosine [$\tan(A) = \sin(A)/\cos(A)$]
Cotangent: ratio of cosine to sine [$\cot(A) = \cos(A)/\sin(A)$]
Secant: reciprocal of cosine [$\sec(A) = 1/\cos(A)$]
Cosecant: reciprocal of sine [$\csc(A) = 1/\sin(A)$]

We can use these definitions—with values determined by the sine and cosine functions we have constructed—to derive all the trigonometric identities we would otherwise have to memorize. For instance, who among us remembers that $\tan^2(A) + 1 = \sec^2(A)$? Fortunately, we don't need to.

If cosine and sine provide the x- and y-coordinates (respectively) on the unit circle, with its radius of length 1, then the Pythagorean theorem relates these values as follows: $\sin^2(A) + \cos^2(A) = 1$. We can use this relationship and the definitions listed earlier to derive all the trigonometric identities. In particular, assuming $\cos(x) \neq 0$, we can divide both sides of the equation by $\cos^2(x)$ to get $\tan^2(A) + 1 = \sec^2(A)$.

We might now have an appreciation for the two special triangles: 30–60–90 and 45–45–90 (aka $\pi/6$–$\pi/3$–$\pi/2$ and $\pi/4$–$\pi/4$–$\pi/2$). They provide benchmark values for constructing sine and cosine, which, in turn, determine the values for all other trig functions. However, in that regard, they aren't alone.

ANOTHER SPECIAL TRIANGLE:
A GOLDEN ONE

We saw in Chapter 8 that Descartes invented the Cartesian plane to solve problems posed centuries before by the ancient Greek mathematician Pappus. Beyond these Pappus problems, the ancient Greeks challenged mathematicians to solve a myriad of construction problems. Those challenges eventually led to the development of whole new branches of mathematics, including abstract algebra.[6]

Recall from Chapter 5 (Table 5.1), Plato's rules for geometric construction. They describe how to produce a line through a pair of points using a straightedge and how to produce a circle centered at one point and passing through another point using a compass. Starting with a pair of points 1 unit apart, all other lengths must be constructed as intersections of lines and/or circles.

Which regular polygons can we construct? What lengths can we construct[7]? Examining Figure 10.1 once more, we can already see that these two questions go hand in hand. The construction of the equilateral triangle, square, and regular hexagon corresponds to constructing points C, B, and A, respectively, which also correspond to the following lengths: $\sin(\pi/3) = \sqrt{3}/2$, $\sin(\pi/4) = \sqrt{2}/2$, and $\sin(\pi/6) = 1/2$. In general, if we can construct a regular n-gon, we can also construct sine and cosine of π/n and vice versa. This is because those values of sine and cosine are just coordinates for one of the vertices of the n-gon.

To construct a mathematical object, we need to understand the relationships that define it. In the case of the regular pentagon—as with the equilateral triangle, the square, and the hexagon—we find special relationships among its angles. By sketching these relationships, as shown in Figure 10.4, and breaking them down, we can reverse-engineer the construction. Specifically, we can take half of the central angle for the regular pentagon, $\pi/5$, and use it to produce isosceles triangle AOB. Note that segment AB, of length s, would be the side of a regular decagon rather than a regular pentagon, but as we will see, if we can construct one, we can easily construct the other.

There is a special relationship among the angles of triangle AOB. Because angle AOB measures $\pi/5$, because triangle AOB is isosceles, and because the angle measures in any triangle sum to π, the remaining two angles in triangle AOB will measure $2\pi/5$. That means when we bisect angle ABO, as shown, with segment BC, we get similar triangle ABC. We can use this relationship and the other relationships shown in Figure 10.4 to determine that segment OC has length s, and segment CA has length s^2. Thus, we have $s + s^2 = 1$. Using the quadratic formula (as derived in Chapter 6), we can then determine that $s = (-1 + \sqrt{5})/2$.

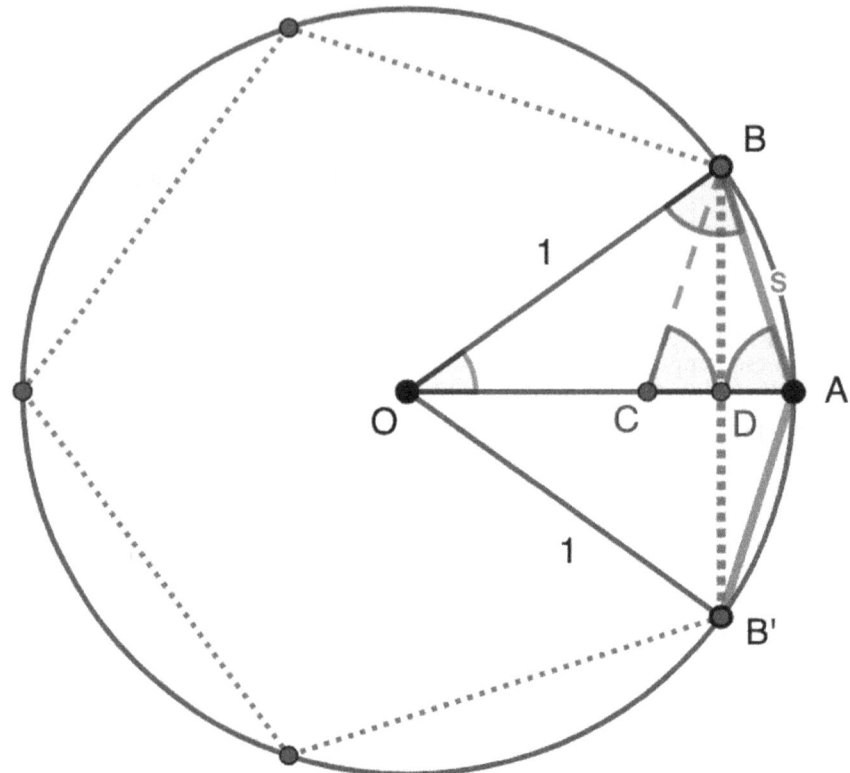

Figure 10.4 Sketching pentagonal relationships.

Reflection

Think about the claims above. How does the similarity between triangles AOB and ABC determine that segment OC has length s and segment CA has length s²? Conversely, if you could construct segment OC (on radius OA) with length s, how might you use it to construct a regular pentagon?

The value of s might look familiar if you have heard of the golden ratio. Usually, the golden ratio refers to the ratio between the length and width of a golden rectangle, which has a special geometric property,[8] and usually we express it as $(1+\sqrt{5})/2$. In that case, it is the solution to $s \times (s - 1) = 1$: a number, s, whose reciprocal is itself minus 1. In our case, we have $s \times (s + 1) = 1$: a number, s, whose reciprocal is itself plus 1. In other words, we are considering the reciprocal of the golden ratio. It is the length we can cut off segment OA so that the remaining segment, CA, is to the cutoff segment as that segment is to the whole: $1/s = s/(1 - s)$.

This is the critical relationship that renders the regular pentagon (and decagon) constructible. If we can construct s, we can construct the regular pentagon.

So how do we construct $(-1 + \sqrt{5})/2$ using Plato's three rules? We might start with the ominous part: $\sqrt{5}$. As we saw in Chapter 3, Thales provided a method for constructing the square root of any positive integer. So here's a plan of action:

1. Given a segment of length 1, we can extend it to a segment of length 6, using circles of radius 1 (Plato's second rule).
2. We can also construct a circle of radius 3 from the midpoint of that segment, passing through each end of the segment.
3. We can construct a line perpendicular to the segment and passing through the point that is five units along the segment. Note that, because constructing perpendicular lines is not directly allowed under Plato's three rules, we need to show how we can accomplish this by coordinating the actions the rules do allow (rotating out circles and sweeping out lines).
4. We can construct the points of intersection between the perpendicular line and the circle of radius 3. We now have the situation Thales described and should be able to argue that one of the segments we have produced has length $\sqrt{5}$.
5. Now that we have a segment of length $\sqrt{5}$, we might subtract a length of 1 from it and then bisect this difference, thus producing $(-1 + \sqrt{5})/2$.

We leave this construction, as well as the construction of the pentagon, as an activity for the end of the chapter, but Figure 10.5 provides illustrations for how these constructions might proceed.

The left side of Figure 10.5 illustrates the first few steps in the construction of s, as described earlier. The right side of Figure 10.5 shows the unit circle, with its radius, OA, partitioned into segments of length s (segment OC) and $1 - s$ (segment CA). Comparing this image to the sketch shown in Figure 10.4, you might imagine how to continue the construction of the regular pentagon (and decagon). With the construction of the regular pentagon, we get a new special triangle: DOB (see Figure 10.4). Like the other special triangles, it is a right triangle, 36–54–90 ($\pi/5$–$3\pi/10$–$\pi/2$), with constructible side lengths. Specifically, OD has length $s + s^2/2$, which means $\cos(\pi/5) = (1 + \sqrt{5})/4$, and we can determine the length of BD, or $\sin(\pi/5)$, using the Pythagorean theorem.

As we saw in Chapter 5, Plato's rules for geometric construction provided the basis for Euclid's postulates—the axioms for Euclidean geometry. These rules, then, were an initial attempt to formalize the kinds of actions we perform to construct mathematical objects. The flourishing of constructions and theorems found in Euclid's *Elements* foretells the proliferation of mathematics you can expect to find through your own mental actions. However, the coordination of those actions also constrains what is possible. Not everything is allowed into the realm of logico-mathematical operations and, likewise, not everything is constructible.

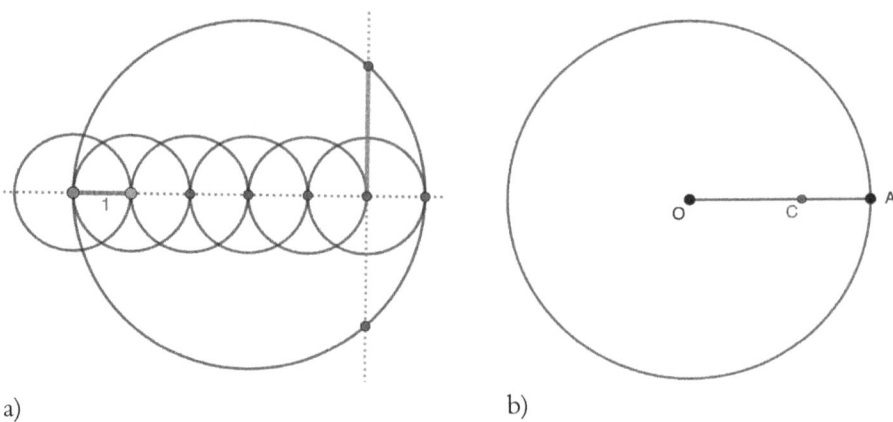

a)   b)

Figure 10.5 Constructing the pentagon.

MORE SPECIAL TRIANGLES

We have argued that constructing special triangles amounts to constructing regular n-gons, and so far, we have ideas for constructing regular n-gons when n=3, 4, 5, 6, or 10. We can also construct polygons with double those numbers of sides by bisecting their central angles. For example, looking at the pentagon sketched in Figure 10.4, we see that we could bisect the pentagon's central angle BOB' to construct the decagon. Likewise, we could bisect angle BOB' in Figure 10.1 to construct an octagon from the square. We could continue these bisections to construct the 20-gon and the 16-gon, the 40-gon and the 32-gon, and so on. In general, if a regular n-gon is constructible, so is any regular $(2^k n)$-gon, where k is any positive integer. Thus, the real problem in determining which regular polygons we can construct is to determine which odd polygons we can construct, and so far, we have only the 3-gon (the equilateral triangle) and the 5-gon (the regular pentagon).

For 2,000 years, beginning in ancient Greece, geometers tried to construct the regular 7-gon (heptagon) using Plato's three rules but to no avail. In the meantime, Gauss found a way to construct the regular 17-gon, which would seem considerably more difficult. As it turns out, the heptagon is not constructible at all. So, what has the 17-gon got that the 7-gon's not got? It's a Fermat prime.

Fermat numbers are of the form $2^{(2^n)} + 1$, where $n \geq 0$. The first five Fermat numbers are prime numbers (numbers whose only divisors, or factors, are 1 and themselves) and are known as Fermat primes.[9] For example, when n = 3, the Fermat number is $2^{(2^3)} + 1 = 257$, which is prime. However, when n = 5 (or, as far as we know, any number $n \geq 5$), the Fermat number is not prime; 4,294,967,297 is divisible by 641. Notice, in Table 10.2, that 17 makes the list of Fermat primes, but 7 does not.

Table 10.2 Fermat numbers and Fermat primes (in bold)

n	F_n
0	$2^1 + 1 = 3$
1	$2^2 + 1 = 5$
2	$2^4 + 1 = 17$
3	$2^8 + 1 = 257$
4	$2^{16} + 1 = 65{,}537$
5	$2^{32} + 1 = 4{,}294{,}967{,}297$

After proving that a regular 17-gon could be constructed with straightedge and compass (i.e., following Plato's rules), Gauss conjectured that a regular n-gon was constructible if, and only if, n satisfied one of the following conditions:

1. n is a Fermat prime. We have already constructed the 3-gon (equilateral triangle) and 5-gon (regular pentagon). We could also construct the regular 17-gon, 257-gon, and the 65,537-gon, with increasing complexity. At least for now, the list ends there. 65,537 is the largest Fermat number determined to be prime.
2. n is a product of distinct Fermat primes. For example, we could construct the regular 15-gon and 51-gon, because 15 = 5 × 3 and 51 = 17 × 3. However, we can't construct the regular 9-gon and 75-gon because their Fermat prime factors are not distinct; they repeat (9 = 3 × 3 and 75 = 3 × 5 × 5).
3. n a power of 2, or a power of 2 times the numbers from case 1 or case 2. For example, we know how to construct the square (n = 2^2) and the regular octagon (2^3). We also know that, given a constructible n-gon, such as the pentagon (n = 5), we can construct a regular 2n-gon, 4n-gon, or in general, a regular 2^kn-gon, by bisecting angles.

Just as the square includes a 45–45–90 triangle and the regular hexagon includes a 30–60–90 triangle, every one of these constructible regular polygons comes complete with its own special triangle. Each of those triangles has constructible lengths, so we can construct the sine and cosine for their acute angles like we did for $\pi/5$ (36 degrees) using the special triangle (the 36–54–90 triangle) within the pentagon (see Figure 10.4). Thus, we can continue generating values for sine and cosine by continuing to construct regular polygons.

Note that cases 1 and 2, earlier, describe all the constructible regular n-gons, where n is odd. Because we know only five Fermat primes, there are only five odd constructible regular polygons that fit case 1, and only 26 more than fit case 2. Would you believe that we can find all these values of n lying in Pingala's triangle ("Pascal's triangle")?

THE MYSTIC TRIANGLE

Recall from the Introduction, that we can generate Pingala's triangle, starting from 1s, by summing pairs of numbers. For example, the number 6 in Table 10.3 is the sum of the 3 above it and the 3 above and to the left of it. We can see lots of patterns in the triangle, but only if we have constructed them. For example, you can readily see the natural numbers in the second column, but you won't recognize the triangular numbers in the third column unless you have cognized them first.

Suppose now, we remove the first row from the triangle and record the remaining numbers as 0s or 1s, depending on whether they are even (0) or odd (1). Then we get the numbers in Table 10.4, which also includes a header row on top and

Table 10.3 Pingala's triangle

1								
1	1							
1	2	1						
1	3	3	1					
1	4	6	4	1				
1	5	10	10	5	1			
1	6	15	20	15	6	1		
1	7	21	35	35	21	7	1	
1	8	28	56	70	56	28	8	1

Table 10.4 The first eight constructible n-gons, where n is odd (Fermat primes in bold)

	1	2	4	8	16	32	64	128	256
3	1	1							
5	1	0	1						
15	1	1	1	1					
17	1	0	0	0	1				
51	1	1	0	0	1	1			
85	1	0	1	0	1	0	1		
255	1	1	1	1	1	1	1	1	
257	1	0	0	0	0	0	0	0	1

a leading column on the left. The header lists powers of 2: $2^0 = 1$, $2^1 = 2$, $2^2 = 4$, and so on. We can use those values to read each row as a binary number. The first row, 1 1, would have one 1 and one 2, which adds to 3 (as recorded in the leading column). The second row would have one 1, zero 2s, and one 4, which adds to 5, and the third row would have one 1, one 2, one 4, and one 8, which adds to 15. These three numbers—3, 5, and 15—represent the first three constructible regular polygons with an odd number of sides. If we extended Table 10.4 to include 31 rows, we could identify the 31 known constructible regular polygons with an odd number of sides.[10]

Findings like this one lead us to believe that mathematics holds a mystical power beyond human understanding—that mathematics lies in wait for us to discover some of its magical secrets while perpetually hiding others. Many mathematicians have felt this way throughout history, but this belief belies the common basis of all mathematical objects. Our own role in construction becomes obscured by the elegance of its polished results.[11]

Reflection

Why do Fermat primes appear within the modified version of Pingala's triangle (Table 10.4), in rows 1, 2, 4, and 8?

Surprising connections between constructible polygons, Fermat primes, and Pingala's triangle raise interesting questions like the one asked earlier. Diving into these questions deeply invariably returns us to the mental actions that define shapes and numbers. Here, we outline answers to some of those questions. A complete and formal investigation would bring us into the realm of Galois theory and beyond the scope of this book. However, as we trace some of the salient mental actions, we sense a psychological foundation for mathematics, even at the most complex levels.

Why do the first 31 rows of Pingala's triangle generate the five Fermat primes and the products of distinct Fermat primes?

Recall that Fermat primes are primes of the form $2^{(2^n)} + 1$. Looking at the header row in Table 10.4, we see that every row of the modified triangle includes a 1 (on the left side) and a highest power of 2 (on the right side). This is because the construction of Pingala's triangle begins with 1s along each side. So, a row in the modified triangle will represent a Fermat prime whenever the exponent for this highest power of 2 is also a power of 2 (e.g., 2^1, 2^2, 2^4, and 2^8), and when the middle terms are all 0. The first row is of this form (with the absence of middle terms), and it is a Fermat prime: $2^1 + 1 = 3$.

As we just noted, every row in the modified triangle (Table 10.4) begins and ends with a 1, representing 1 plus some power of 2. The terms in the middle of each row are determined by adding two terms from the row above. In the case of the second row in Table 10.4, this sum is $1 + 1 = 2$, which is even, so it appears as 0. Also note that, because each row in the modified triangle includes one more term than the row before it, its highest power of 2 will increase by 1. Specifically, in moving from the second row to the third row, this highest power of 2 will increase from 2^1 to 2^2. Once again, the exponent (2) is a power of 2, so we have another Fermat prime.

The third row does not represent a Fermat prime because its highest power of 2 is 2^3, and 3 is not a power of 2 (i.e., it is not of the form 2^{2^n}). Also, it has nonzero terms in the middle. However, you might recognize the number this third row represents as the product of distinct Fermat primes. Specifically, 1,111, in base-2, represents $8 + 4 + 2 + 1 = 15$, in base-10, and this is the product of Fermat primes 3 and 5. If we consider the distributive property of multiplication (discussed in Chapter 1) and how multiplication works in base-2 (returning to an idea about the binary system introduced at the end of Chapter 4), we can see how the modified triangle produces every possible product of distinct Fermat primes.

In the case at hand, we have Fermat primes 3 and 5, which we represent in base-2 as 11 and 101. In binary (base-2) arithmetic, this product is $11 \times 101 = (10 + 1) \times (100 + 1) = 1{,}000 + 100 + 10 + 1 = 1{,}111$. Now, we need to show that we get this same result in constructing the third row of the modified triangle. Like all the other rows, the third row will start and end in a 1. To get the middle terms in that row, we look to the row above (the second row), which is 101. The sum of the first two terms, $1 + 0$, will generate one of the middle terms in the third row, and the sum of the last two terms, $0 + 1$, will generate the other middle term in the third row. So, we do indeed generate 1,111 in the third row, showing that the third row represents the product of the Fermat primes represented by the first two rows ($3 \times 5 = 15$). There's more work to be done to show why the modified triangle generates *all* products of distinct Fermat primes (and only Fermat primes or products of distinct Fermat primes, for the first 31 rows), but we can already see how it breaks down to mental actions, like distributing units (the powers of 2 within binary arithmetic), as represented within Pingala's triangle.

Why are Fermat primes the only prime numbers, p, for which we can construct regular p-gons?

Keep in mind that Plato's method of geometric construction was not ordained by God. It has roots in the mundane activity of drawing lines and circles. The activity became interesting to ancient Greeks when they began to wonder what mathematical objects they might construct from such simple rules. The construction of regular polygons became a sort of game to play. The Greeks elevated the game to the ideal world of mathematics exactly when they internalized it as a coordination of mental actions rather than the sketching of figures in sandboxes.

However natural the rules of the game might seem, they could have been different. For example, if the Greeks allowed for the use of a string to measure off the circumference of a circle and lay it out in a line, they could construct 2π as a length. Thus, they could square a circle, and the set of constructible polygons would change. In other words, if we changed the mental actions underlying geometric construction—from sweeping out lines and rotating distances around points to draw circles, to some other set of mental actions, like straightening circles—the set of constructible figures would change too. In particular, constructible regular p-gons would lose their mysterious connection to Fermat primes.

The construction of regular p-gons inscribed within the unit circle coincides with the construction of lengths, especially the x- and y-coordinates on the unit circle. We construct trigonometric functions by coordinating these covarying quantities with each other and the central angle. In Chapter 9, we saw how Descartes could describe lines and circles with equations that represented the covariation of two quantities—two distances from a pair of axes, which serve as a reference frame for all other points in the plane. Specifically, the Pythagorean theorem describes the invariant relationship between the x- and y-coordinates of points on the unit circle: $x^2 + y^2 = 1$. So, we can identify the coordinates for points of intersection between lines and circles by taking square roots. Case in point, that's how we constructed lengths for the coordinates $\cos(\pi/5)$ and $\sin(\pi/5)$ on the regular pentagon.

Figure 10.6 illustrates the regular pentagon a little differently. Taking advantage of the ideas investigated in Chapter 8, we consider the five vertices of the

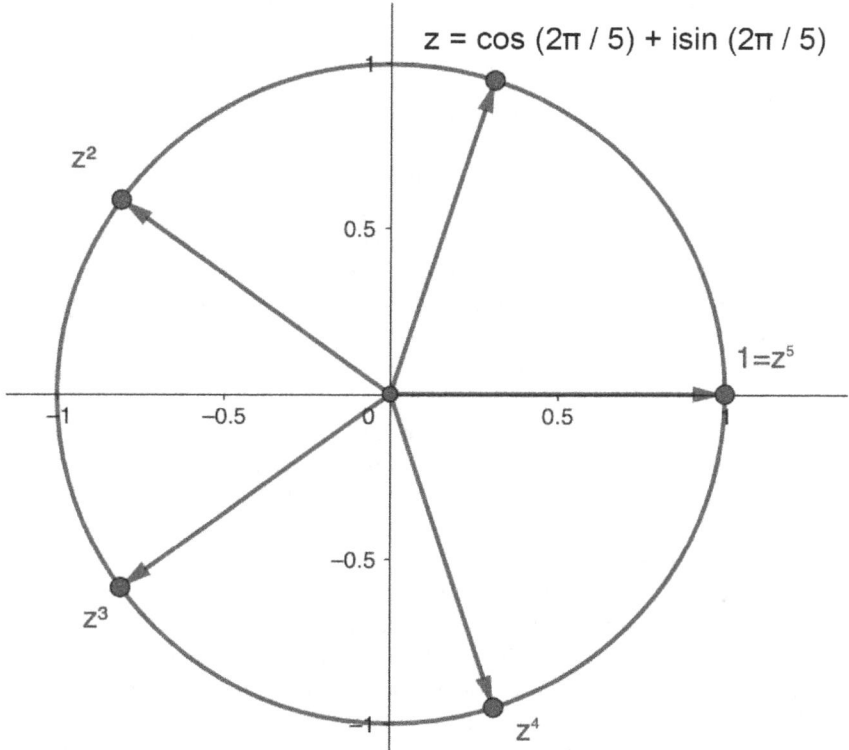

Figure 10.6 The regular pentagon in the complex plane.

regular pentagon as vectors in the complex plane. We place one vertex at 1, then the next vertex is represented by the vector $z = \cos(2\pi/5) + i\sin(2\pi/5)$, and the remaining three vertices are just powers of z: z^2, z^3, and z^4. To see why these powers work out the way they do, we can rely on the double angle formulas for sine and cosine, which you can derive in Activity 4. For example, $z^2 = [\cos(2\pi/5) + i\sin(2\pi/5)] \times [\cos(2\pi/5) + i\sin(2\pi/5)] = \cos^2(2\pi/5) - \sin^2(2\pi/5) + 2i\cos(2\pi/5)\sin(2\pi/5)$. By the double angle formulas, this is just $\cos(4\pi/5) + i\sin(4\pi/5)$.

Now, considering 1, z, z^2, z^3, and z^4 as vectors, and relying on the symmetry of the pentagon, you might see that $z^4 + z^3 + z^2 + z + 1 = 0$ (because the vectors start from 0 and end in symmetric positions around the circle, they balance out to 0, just as 1 and -1 do). So, we have a fourth-degree polynomial whose roots are vertices of our pentagon. Moreover, because 5 is prime, we can start from any of those four roots and generate all the other roots by repeatedly squaring this first root. For example, starting from z^3, we get $(z^3)^2 = z^6 = z^5 z = z$; squaring that term we get z^2, and squaring that term we get z^4. With the double angle formulas, we see that all the roots appear in the same field extension.

The key here is that the degree of the polynomial is a power of 2 and that the roots form a cycle (as illustrated in Figure 10.6) so that we can generate all the roots from any one of the roots. In the case of the equilateral triangle, we have $z^2 + z + 1 = 0$, and we can use the quadratic formula to find constructible values for $\cos(2\pi/3)$ and $\sin(2\pi/3)$; namely $z = -\dfrac{1}{2} \pm \dfrac{\sqrt{3}}{2}$, so $\cos(2\pi/5) = -1/2$ and $\sin(2\pi/5) = \sqrt{3}/2$. In the case of the regular pentagon, we could use the quartic formula (discussed in Chapter 6), which involves taking square roots of square roots and generates constructible values for $\cos(2\pi/5)$ and $\sin(2\pi/5)$. For the 17-gon, we have a polynomial of degree 16, or 2^4. We have no formula for solving all degree-16 polynomials, but we can find the roots of this one by taking square roots of square roots of square roots of square roots.[12]

In general, for a regular p-gon to be constructible, we need p-1 to be some power of 2, so that we can derive z through a sequence of nested square roots and, therefore, can construct its coordinates as intersections of lines and circles. To complete our sketch of the argument about Fermat primes and constructible polygons, we would also need to show that, in order for $2^k + 1$ to be prime (i.e., for $p - 1$ to be a power of 2), k must also be a power of 2. This claim, too, follows from the distribution of units.[13]

Why are distinct products of distinct Fermat primes the only other odd values of n for which we can construct n-gons?

We have already constructed the equilateral triangle and the regular pentagon. How we might combine those constructions to construct the regular 15-gon? Figure 10.7 illustrates the situation.

Note that the construction of the equilateral triangle includes the point $(\cos(2\pi/3), \sin(2\pi/3))$, and that the construction of the regular pentagon includes

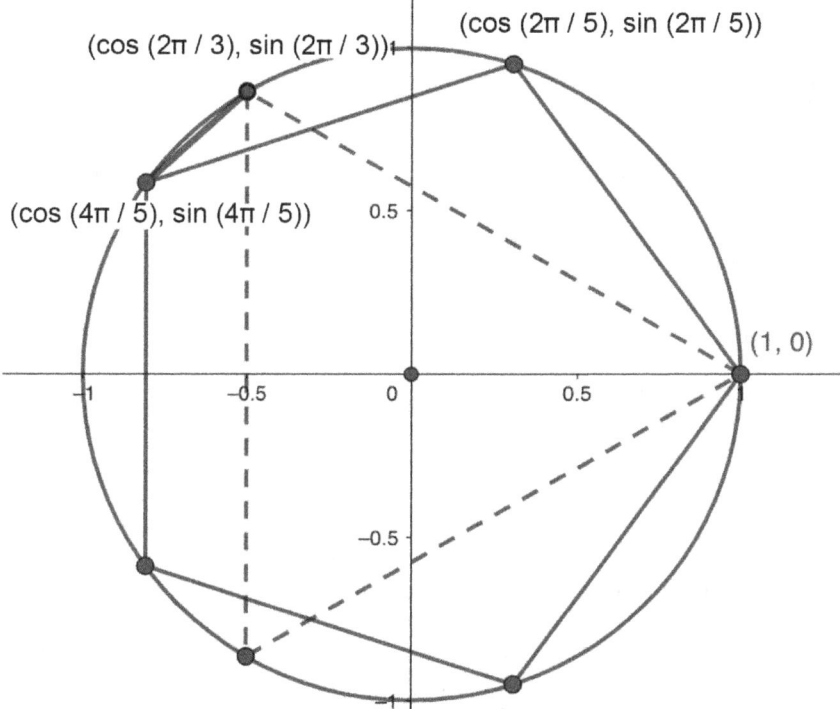

Figure 10.7 Multiplying n-gons.

the point $(\cos(4\pi/5), \sin(4\pi/5)$. Partitioning the angles $2\pi/3$ and $4\pi/5$ into 15ths, we have $2\pi/3 = 10\pi/15$ and $4\pi/5 = 12\pi/15$. These are two adjacent vertices on the 15-gon: $5(2\pi/15)$ and $6(2\pi/15)$. Thus, connecting those two vertices forms a side of the 15-gon, and we can construct all the other sides by swinging circles with this side length.

Reflection

Can you generalize this argument to construct any $(p \times q)$-gon given a regular p-gon and a regular q-gon, where p and q are different prime numbers? Why doesn't it work with p and q are the same prime number?

Once again, we have omitted many details but nonetheless indicate the psychological basis for mathematics. A more rigorous explanation would benefit from Galois theory and abstract algebra. In abstract algebra, we invent new notational devices for symbolizing mental actions, with computational methods for simulating their coordination. We saw simpler cases of this in Chapter 6, regarding the use of high school algebra. We can think about abstract algebra as a further abstraction whereby we can study whole systems of operating:

> for it is only at the end of a sufficiently long series of reflective abstractions that the subject discovers the most profound characteristic of operations: that of being connected together in structures which have their own laws of totality. That is why we had to wait for E. Galois to discover the group concept which Viete or Descartes, nevertheless, constantly used unconsciously in their algebra.[14]

Galois theory, and abstract algebra more generally, brings to our attention the regulation of our own mental activity. In Chapter 3, we considered Klein's Erlangen program of classifying geometries based on groups of transformations. We extended the program to demonstrate the power of our own mental actions in constructing and transforming space. These transformations are characterized and regulated by their composability and reversibility—two of the defining characteristics of algebraic groups. Abstract algebra, then, becomes the study of regulations of our own mental activity, especially the coordination of our own mental actions.

Formally speaking, abstract algebra did not become a branch of mathematics until the 1800s, but we can trace its roots to some of the earliest conceptions of number and geometry. For example, fractions form a group under the operation of multiplication (the splitting group, discussed in Chapter 4). When we extend them in the negative direction and include the operation of addition, they form the field of rational numbers (as discussed at the end of Chapter 8). We can extend that field to include other constructible numbers, the same way we extended rational numbers to the field of (rational) complex numbers. For instance, in constructing the equilateral triangle, we constructed $\sqrt{3}/2$, along with all additive and multiplicative combinations of $\sqrt{3}$ with other numbers in the field: $a + b\sqrt{3}$, where a and b are rational numbers.

SUMMARY

We began this chapter with a question about special triangles and concluded it with a classification of constructible polygons. The short answer to our question is that we can use special triangles to support our constructions of trigonometric functions.[15] The longer answer is that each constructible regular polygon comes with its own special triangle, adding more and more exact values to Table 10.1 and Figure 10.2. These values serve as benchmarks in our construction of trigonometric functions, through covariations of angles and lengths on the unit circle. Ultimately, a simple question led us down a path of mathematical proliferation so complex that we might forget we are the ones laying out the path through our own mental activity.

We often think about mathematics as a growing list of distinct branches that correspond with course titles, such as Number Theory, Geometry, Complex Analysis, and Abstract Algebra. Then we marvel at the idea that complex analysis can be used to prove theorems about prime numbers; such was the case with Fermat's last theorem.[16] As we have seen over and over in this book and throughout the history of mathematics, complexities do manifest themselves but not by themselves. They occur through our own generalizations, coordinations, and regulations in response to problems that interest us. After all, if Plato had chosen different rules for geometric construction, the constructible polygons—along with their intimate connections to Fermat primes and Pingala's triangle—could be entirely different. But we are not bound by Plato's rules. We can play our own mathematical games (as with the invention of non-Euclidean geometries, which plays by different rules) and pursue our own mathematical interests. Therein, we are bound only by the reversibility and composability of our own mental actions.

Activities

Activity 1: Recall Activity 1 from Chapter 5, which involved constructing a square using Plato's three rules of construction. The first two activities in this chapter involve additional constructions. Starting from the unit circle, construct an equilateral triangle inscribed within it, then show how to bisect one of its central angles. The result should look something like the inscribed triangle in Figure 10.1.

Activity 2: Construct the golden ratio and use it to construct the regular pentagon. It might help to refer to the left and right sides of Figure 10.5 respectively, for each construction.

Activity 3: Referring to Table 10.1 and Figure 10.2, what value would you expect $\cos(\pi/5)$ to have? Now compute its value based on the construction of the regular pentagon. Do the same thing for $\cos(\pi/8)$ and the construction of the regular octagon.

Activity 4: Figure 10.8 shows a unit circle with triangle ABD rotated by angle a, to congruent triangle ACE. Thus, angle DAC measures 2a. Coordinates on the unit circle, at points B and C, are labeled accordingly. Derive various side lengths and use them to prove the double angle formulas for sine and cosine: $\cos(2a) = \cos^2(a) - \sin^2(a)$ and $\sin(2a) = 2\cos(a)\sin(a)$. The dashed lines may be useful.

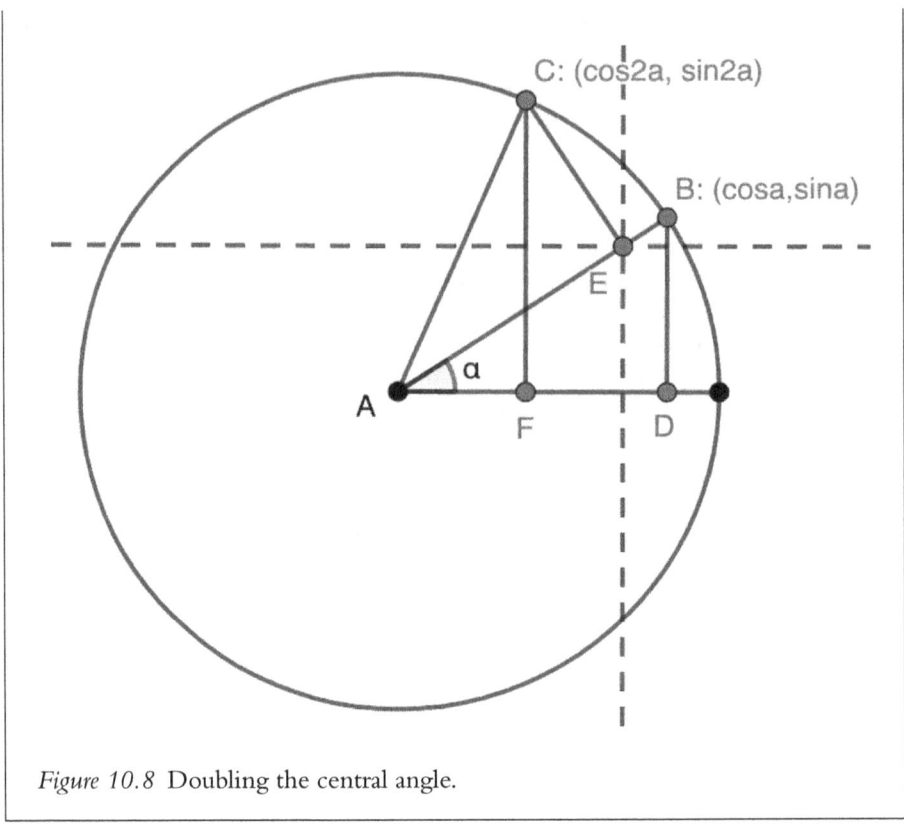

Figure 10.8 Doubling the central angle.

NOTES

1. A mathematics education researcher and friend of mine, Keith Leatham, posed this question. Like all good questions, it's one that has stuck with me, driving me to investigate it deeper over time. Subsequent conversations with colleagues in my department, Bud Brown and Nick Robbins, have helped me do so.

2. This activity is described in further detail by Moore and LaForest (2014) in the National Council of Teachers of Mathematics journal, *The Mathematics Teacher*—an excellent resource for teachers in planning lessons.

3. In the *Republic*, Plato (1952) introduces Socrates as a wise character to explain his philosophy using the Socratic method of asking questions to his pupils so that they might arrive at the truth. Later in the book, Socrates asks similar questions about numbers, chastising his pupils for focusing on figures rather than ideals: "arithmetic has a very great and elevating effect, compelling the soul to reason about abstract number and rebelling against the introduction of visible or tangible objects into the argument" (Book VII, 525c).

4. Note that "sine over cosine" gives the same ratio as the "opposite over adjacent" definition you might have learned in high school with reference to right triangles, wherein sine is "opposite over hypotenuse" and cosine is "adjacent over hypotenuse."

5.   See Barrera's (2014) article in the *Mathematics Teacher* for a clear and detailed discussion of trigonometric functions as lengths and using those lengths to determine angles associated with the inverse functions. The article includes suggestions for teaching trigonometric functions to high school students.

6.   Galois theory, named for the 19th-century French mathematician Éveriste Galois, allows for elegant solutions to all kinds of construction problems. Similar to the way the Erlangen program (Chapter 3) used groups to classify geometries, Galois theory uses groups to describe chains of field extensions. We can use it to determine which numbers can be geometrically constructed and which roots can be algebraically derived. For example, we can derive a formula for finding roots of quadratic equations (as we did in Chapter 6) and even cubic and quartic equations, but we can't derive a formula for polynomials of degree five and higher.

7.   Greeks treated numbers as lengths and so it made sense for them to ask questions about constructible numbers.

8.   See Brown (1976) for more details on the Golden rectangle, its construction, and its connections to science and nature.

9.   Fermat invented Fermat numbers in a fruitless attempt to find a prime number generator—a rule that generates an infinite number of prime numbers. Such a thing would be of great interest to cryptologists today.

10.   John Conway and Richard Guy (1998) wrote about this relationship between Pascal's/Pingala's triangle and constructible polygons in their *Book of Numbers*, which includes numerous examples of unexpected connections.

11.   Philosopher Imre Lakatos (1976) wrote *Proofs and Refutations* as a challenge to formalist views of mathematics and as a partial rebuke of the Platonic notion that mathematical objects existed prior to the human labor of creating them. Interestingly, he wrote in the same Socratic style as Plato did.

12.   The distinction here relies on the irreducibility of this polynomial and Galois theory.

13.   When n is odd, $(x^n + 1) = (x + 1)(x^{n-1} - x^{n-2} - \ldots -x+1)$. Suppose k has an odd factor. In other words, suppose $k = 2^j n$ for some odd number n. If we set $x = 2^j$, we have that $2^k + 1$ is divisible by $2^j + 1$, and so it's not prime. For example, if $k = 6$, $2^k + 1$ is divisible by 5.

14.   This quote comes from Beth and Piaget (1966, p. 294). Evert Beth was a logician who studied the foundations of mathematics.

15.   It's also worth noting that the 30–60–90 and 45–45–90 triangles are special/unique in the sense that they are the only ones that give a rational value for sine, cosine, or tangent. This observation was made by my friend, mentor, and colleague, Bud Brown.

16.   Fermat posited his "last theorem" in 1637, claiming (but not proving) that $x^n + y^n = z^n$ has no solutions for nonzero integers x, y, and z when n is an integer greater than 2. The conjecture itself is surprising given that there are plenty of solutions when n = 2; these relate to right triangles and are known as Pythagorean triples, such as when $x = 3, y = 4$, and $z = 5$. The proof (offered three and a half centuries later, by Andrew Wiles) is surprising because it relies on elliptical curves in the complex plane (Singh, 1997). Indeed, much of the recent progress in solving open conjectures in number theory has likewise relied on complex analysis, especially use of the Reimann zeta function.

REFERENCES

Barrera, A. (2014). Unit circles and inverse trigonometric functions. *The Mathematics Teacher*, *108*(2), 114–119.

Beth, E. W., & Piaget, J. (1966). *Mathematical epistemology and psychology* (W. Mays, Trans.). Dordrecht: D. Riedel.

Brown, S. I. (1976). From the golden rectangle and Fibonacci to pedagogy and problem posing. *The Mathematics Teacher, 69*(3), 180–188.

Conway, J. H., & Guy, R. (1998). *The book of numbers.* New York: Springer Verlag.

Lakatos, I. (1976). *Proofs and refutations: The logic of mathematical discovery.* Cambridge: Cambridge University Press.

Moore, K. C., & LaForest, K. R. (2014). The circle approach to trigonometry. *The Mathematics Teacher, 107*(8), 616–623.

Plato. (1952). *The republic* (B. Jowett, Trans.). Chicago: William Benton.

Singh, S. (1997). *Fermat's enigma: The epic quest to solve the world's greatest mathematical problem.* New York: Walker and Company.

11

Calculus the Old-Fashioned Way

Prior chapters have illustrated the unity of space and number. In Chapter 8, we saw how we could extend numbers into new directions in space. We created integers and complex numbers by introducing mental actions of reflecting and rotating (first used to construct space) into existing number systems. Chapter 10 demonstrated ways that ancient Greek mathematicians constructed numbers in space. The Greeks would represent these constructible numbers as lengths. Placing those lengths on the same line (the continuum) gave the Greeks a kind of number line.

Today, schooled in modern mathematics, we tend to think about the number line as a collection of points. We rely on decimal expansions to describe these points, and we accept that even simple numbers, like 1/3, may require unending expansions: $0.3\bar{3}$. Other numbers, including constructible ones like $\sqrt{2}$ and non-constructible ones like π, have non-terminating decimal expansions that don't even repeat. We accept that we can find an infinite number of numbers between any two real numbers and that there is no number next to 0, $\sqrt{2}$, or any other real number. We might even begrudgingly accept that the non-terminating decimal expansion $0.9\bar{9}$ is equal to 1. What we have done, in accepting real numbers as real, is to replace the intuitive idea of the continuum with a collection of points. Instead of taking the continuum as given—as a line swept out in space—and constructing points on it, we take an uncountable collection of points for granted and claim it constitutes a line.

DOI: 10.4324/9781003181729-12

Reflection

Consider the following arguments that $0.9\overline{9}$ =1. First, if $0.9\overline{9}$ and 1 are different real numbers, there must be some other real number between them, namely their average; so what is that number? Second, if we accept that $0.3\overline{3}$ =1/3, it follows that $0.9\overline{9}$ =1 (just multiply both numbers by 3). Do you find these arguments convincing? How would you respond to them?[1]

This chapter begins with an investigation of real numbers, questioning the sense in which we might call them real. It then picks up on the conversation of continuity, begun in Chapter 9. Concepts of real numbers, the continuum, and continuity provide the foundation for considering instantaneous rates of change—the central concept of calculus. We will investigate this concept the old-fashioned way, as Newton did, with covariation in mind.

GETTING REAL

For the ancient Greeks, the continuum was a continuous line, stretching out without bounds in two opposite directions. It was a space on which to construct points. It existed in their minds as a blank canvas on which to perform their mathematical art. It was not constituted of points any more than a canvas is made up of brush strokes. However, as Greek mathematicians practiced their art, the line become more and more populated with densely packed points. In fact, rational numbers densely populate the continuum so that we can approximate every possible location on it, as closely as we like, by a positive or negative fraction. Even non-constructible numbers, like the cubed root of 2, we can approximate by fractions and, thus, locate them on the continuum.

Reflection

Make a list of fractions close to the cubed root of 2. Separate this list into two sublists: one with fractions a little less than the cubed root of 2 and one with fractions a little greater than the cubed root of 2. Use those lists to find fractions closer and closer to the cubed root of 2.

In generating two lists of fractions relative to a given real number, r (e.g., the cubed root of 2), you are defining the number through a limiting process. You are creating a Dedekind cut between the set of rational numbers less than r and the set of rational numbers greater than r. In this manner, Dedekind cuts define an irrational number as two infinite sets of rational numbers. If you reflect on the situation a bit more, you might realize this is not so different from what we do in determining the decimal expansion for a real number. But if we define real numbers as infinite sets or infinite processes, in what sense are real numbers real?

We can geometrically construct rational numbers and, in some cases, irrational numbers too. Yet, the vast majority of real numbers are not constructible. Worse yet, most of those numbers are not even algebraic: we don't even encounter them in solving algebraic equations like we do the cubed root of 3 (a solution to the equation $x^3 = 2$). They are transcendental (e.g., π, e, and the natural log of 2).[2] They transcend attempts to determine their exact values through finite processes like geometric construction and even algebraic manipulation.

With the invention of calculus in the 17th century and limits in the 19th century, mathematicians began to accept transcendental numbers, and irrational numbers in general, as points that filled out the continuum. At that point, mathematicians began to think of the continuum as an uncountable collection of points. In the 1920s, Brouwer and like-minded "intuitionist" mathematicians and philosophers pushed back on this idea.[3] They reminded us that numbers are products of human thought whose construction must be specified by some sequence of actions. Numbers could not be taken for granted.

In our own construction of numbers so far, we began with whole numbers, which rely primarily on the mental actions of unitizing and iterating. When we included the mental action of partitioning and coordinated it with iterating, we generated fractional units and the splitting group (Chapter 4). When we included the mental action of reflection, we generated numbers in two directions, giving us integers (Chapter 8). Combining those two systems, we achieve the field of rational numbers (both positive and negative fractions). In Chapter 6, we saw how we could extend the geometric construction of number through algebraic manipulation. Now, because some numbers transcend finite processes (geometric or algebraic), we turn to limiting processes, which involve a potentially infinite sequence of actions, like Dedekind cuts and decimal expansions.

In modern mathematics, we rely on a formal definition of limits to define continuity and, thereby, to provide a foundation for calculus. However, calculus was invented about 200 years before limits were formally defined. Both continuity and calculus relied on more intuitive understandings of limiting processes.

CONTINUOUS FUNCTIONS

As noted in Chapter 9, continuity is a property of functions. It describes a relationship between variables. When acting in the world, time is the ever-present variable, and like the continuum, we sweep it out in a line. When we consider other variables, it is natural for us to think about them varying in time. When we refer to a continuous variable, x, we have in mind x(t), with x varying as a function of time, t. We say that this variable is continuous because, as it varies in time, like time, it sweeps out in a line. However, unlike time, which sweeps out uniformly in one direction, this other line might lengthen, shorten, or even move in the opposite direction, from one unit of time to the next. For example, consider the function x(t) = sin(t), discussed in Chapter 10. As t increases from 0, x increases quickly, slows down, reverses directions, decreases more quickly, then slows down and changes directions again, and so on, in a cycle (period) that recurs every 2π units of time (see Figure 11.1).

Here, we restrict our attention to variables that vary on the continuum (formally speaking, we focus on functions that map the real line to the real line). In particular, as t varies in time, x varies within another copy of the continuum. We can see this by focusing on the position of t on the horizontal axis and the position of x on the vertical axis, as the point P, which represents (t, x), moves along the curve (see Figure 11.1). The two variables covary in such a way that between any two moments in time, say, t = 2 and t = 3, x(t) takes on all the values between x(2) and x(3). This property is essentially the intermediate value theorem (IVT).

Reflection

Imagine a rope attached, on either end, to a pair of posts separated by about 10 meters. If you detach the rope, tie it in knots (but do not cut it!), and throw it in a pile somewhere between the two posts, there will be some point on the rope whose distance to each post does not change.[4] Do you believe it? Could you prove it? How might it relate to the IVT?

As illustrated in Figure 11.1, IVT implies that the graph of x(t) will be connected and gapless. This is the property that people rely upon when they describe continuity as "a process that proceeds without gaps or interruptions or sudden changes."[5] This description might work as a definition of continuity in the present

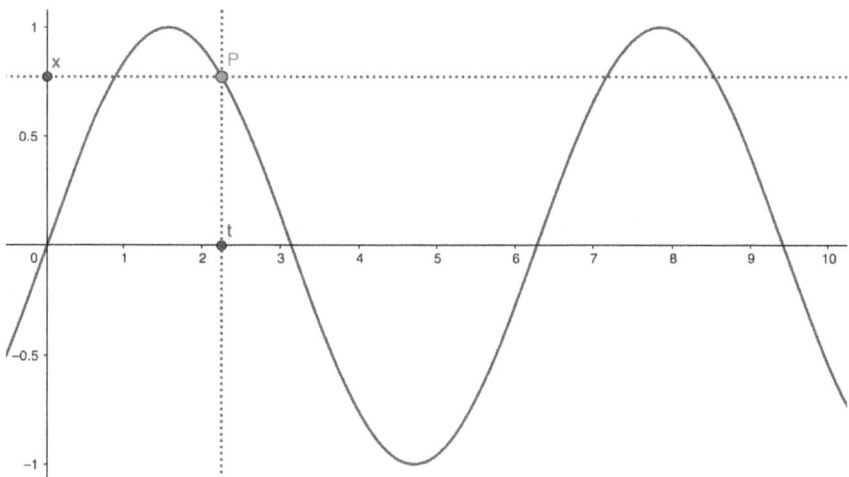

Figure 11.1 Graph of the continuous function x(t) = sin(t).

context of functions that map from the continuum to the continuum. However, it fits better as a property of the graph (a connected graph) than it does as a property of the function itself (a continuous function). In line with our investigations of covariation, begun in Chapter 9, we wish to look beyond the graph and into the mental actions we use to coordinate the two covarying quantities. Moreover, as we will see, Newton's original and intuitive approach to calculus relied on something closer to the formal definition of continuity (the epsilon–delta definition[6]) than the IVT. We can describe the IVT better as a consequence of continuity than as a definition of it.[7]

RATES OF CHANGE

As originally conceived, in the 17th century, calculus was the art of calculating with infinitely small numbers. You can think of these infinitesimals as smaller than any positive real number but bigger than zero. If you don't believe $0.9\overline{9} = 1$, you might think of an infinitesimal as the difference between those two numbers. In fact, the equality of 1 and $0.9\overline{9}$ depends on rejecting the existence of infinitesimal numbers. Arguments against the equality often refer to an infinite sequence of partitions, one-tenth (0.1), partitioned into hundredths (0.01), partitioned into thousandths (0.001), and so on, without end. To represent an infinitesimal as a decimal, we would need to write an infinite number of 0s and append a 1 at the end.

Infinitesimals are nonzero, but also not units because we can't use them to measure quantities. They stand in contrast with even very small fractions and decimals. We can produce arbitrarily small unit fractions, but no matter how small, there is some number of iterations of a unit fraction that will re-reproduce the unit 1 (i.e., we can measure with them). For example, one billion iterations of one-billionth will get us back to 1. In contrast, infinitesimals are infinitely small, so no number of iterations of them will amount to 1. As such, we have to be careful in calculating with them. Newton calculated with these infinitely small quantities to determine instantaneous rates of change: derivatives.

We have considered rates of change within Carlson's framework for covariation (Chapter 9). We considered how one variable changes as another variable increases one unit at a time. At higher levels in Carlson's framework, students might partition that unit and consider how the first variable changes with respect to finer and finer units. Derivatives are the culmination of those efforts to refine rates of change, to determine an instantaneous rate of change.

In calculus class, we traditionally teach derivatives with limits. Here, we consider derivatives as Newton did, by comparing changes in variables at a moment in time. Newton considered variables as varying in time. When he referred to a variable, like x, he had in mind x(t), as discussed in the prior section. He referred to variables like these as "fluents" because they flowed fluidly in time. Newton was interested in how these fluents flow, or vary, at a moment in time.

A moment is an infinitesimally small value of time, like an instant in time. Newton symbolized a moment in time with an "o"; kind of like a zero but not quite. At a given moment, x varies at some rate. Newton referred to this rate as a "fluxion" and symbolized it with an "ẋ." The product ẋo, then, gives the amount by which x varies at a moment in time. Take, for example, $x(t) = t^2$, whose graph is shown in Figure 11.2.

At a moment in time, t varies to become $t + o$, and x covaries to become $x + ẋo$, also known as $x(t + o)$. Because the equation, $x(t) = t^2$, represents this covariation, it works with these new values of t and x, as well as it did for the original values. In other words/symbols, $x + ẋo = (t + o)^2$. If we treat o as if it were a real number (as Newton did) and distribute the terms in $(t + o)(t + o)$, we would get $x + ẋo = t^2 + 2to + o^2$. Because we know $x(t) = t^2$, we can subtract out the terms x and t^2, as they appear on each side of the equation, leaving $ẋo = 2to + o^2$. If we treat o as if it were a unit, we could divide it out, leaving $ẋ = 2t + o$. If we then treat o as 0, we would have $ẋ = 2t$.

How should we interpret this equation, $ẋ = 2t$? What does it tell us about the original equation, $x(t) = t^2$, and its graph (Figure 11.2)? Both the original equation and its graph represent how x and t covary. This new equation represents the rate

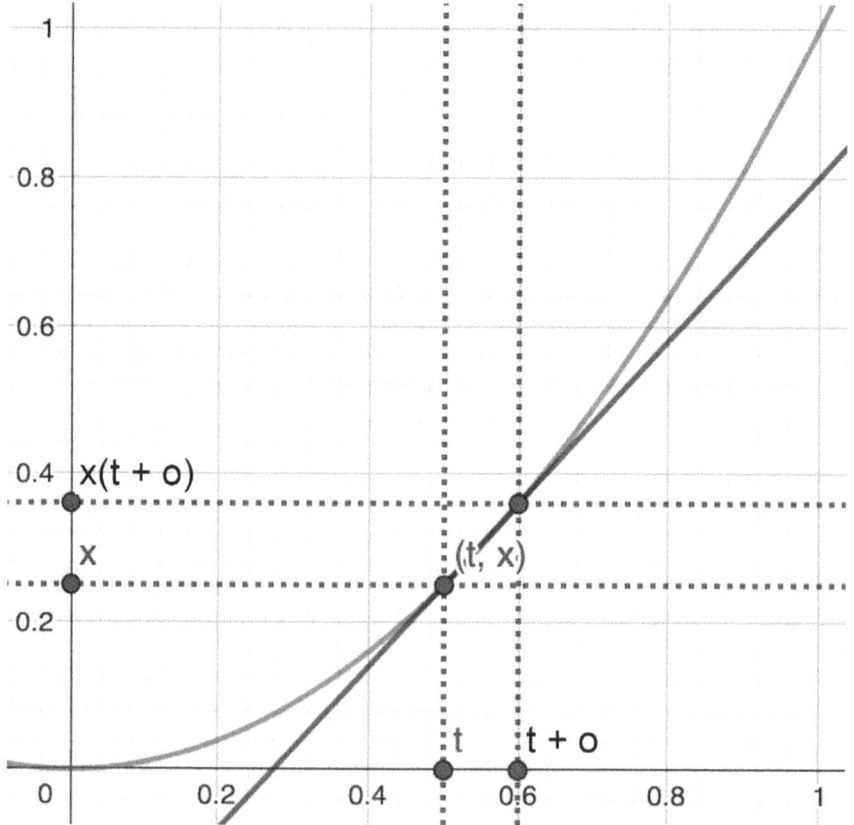

Figure 11.2 Graph of $x(t) = t^2$ and its instantaneous rate of change.

at which x is changing at a moment in time—at time, t, in particular. When t = 0, 2t = 0, so the instantaneous rate of change in x is 0 at that time. When t = 1, 2t = 2, so at that moment in time, x changes at twice the rate t changes. Whereas the original equation relates the values of x and t as they covary, the new equation relates their rates of change.

The graph illustrates these rates of change as slopes of a tangent line. At each point on the curve (parabola), the tangent line touches it at a single point, and its slope approximates the relative rate of change between the variables x and t at that point, (t, x). Notice how the slope of this line changes as we move from the point (0, 0) to the point (1, 1). However, at each point, (t, x), we can describe the slope as 2t.

Reflection

Look back at the hourglass problem from Chapter 9 (see Figure 9.4) wherein we identified an invariant relationship between height and volume. Namely, the rate at which the rate of change of volume varied, in relation to height, was constant. How does that relate to the present situation where we have 2t as a rate of change?

GHOSTS OF DEPARTED QUANTITIES

Newton treated o as a nonzero number when dividing but then treated it as zero when eliminating the term that still contained it.[8] He wanted to have it both ways: o as a unit and o as 0. As discussed in Chapter 8, 0 is the one real number that we can't treat as a unit. No one could deny that Newton's method worked, but they certainly questioned its theoretical foundations. Bishop Berkeley, an 18th-century theologian and mathematician, mocked the method in asking, "What are these fluxions? The velocities of evanescent increments? And what are these same evanescent increments [moments]? They are neither finite quantities, nor quantities infinitely small, nor yet nothing. May we not call them ghosts of departed quantities?" Calculus faced a crisis in foundations that went unsettled until 200 years after Newton (and Leibniz) invented calculus. Finally, in the 19th century, the epsilon–delta definition of limits supplanted the need to calculate with infinitesimals and provided an acceptable foundation for defining continuity, derivatives, and integrals.

Recall from Chapters 1, 4, and 8 that multiplication involves a transformation of units. Consider the case of the product ẋo. If we treat o as if it were a unit, we would have the unit transformation from 1 to o. This would be an extreme version of the kind of unit transformation Carlson suggests in her covariation framework (see Chapter 9). In that framework, we partition the units measuring the independent variable into smaller and smaller units, in order to get finer and finer measures for the rate of change in the other (dependent) variable. For example, consider the left side of Figure 11.3.

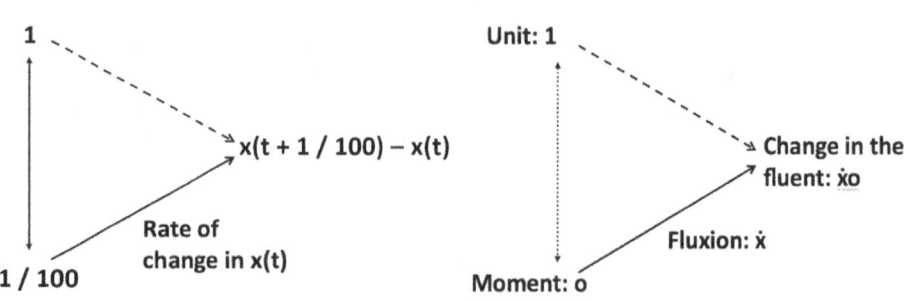

Figure 11.3 The fluxion as a transformation of units.

Source: © Eleanor Norton

Taking time as the independent variable, measured in units of 1 (say, 1 second), we get a rate of change in x(t), which would be x(t + 1) − x(t). However, if we partition the unit of 1 into hundredths, we get a better approximation of the instantaneous rate of change: x(t + 1/100) − x(t). This improved approximation depends on a transformation of units, from units of 1 to units of 1/100. Taken to the extreme, we would transform units of 1 into moments in time, o, giving us an instantaneous rate of change, ẋo. However, o is not a unit because, as an infinitesimal, it can't measure quantities.

In answer to Berkeley's question, Newton wasn't too concerned about the objective status of moments, as evanescent increments, "ghosts of departed quantities," nor as units. He cared only about the instantaneous rate of change in x, as symbolized by the fluxion, ẋ. In the transformation of units shown on the right-hand side of Figure 11.3, it's the action itself (represented by the solid arrow) that matters. Newton didn't need o to be an actual unit because he wasn't using it to measure a quantity; he was using it to measure ẋo, which we might call the ghost of another departed quantity. All he required was that ẋo be some measure of o. In other words, he needed for changes in x(t) to be arbitrarily small whenever changes in time were arbitrarily small. In modern, formalized calculus, this requirement is codified by the epsilon-delta definition of continuity. Thus, we see that Newton's method for calculating derivatives did not appeal to gaplessness or the IVT. Rather, it implicitly relied on the kind of covariation between variables that the modern definition of continuity attempts to capture.

TAKING DERIVATIVES OF PEAKS AND CIRCLES

In high school calculus, we learn to take the derivatives of functions, f(x), wherein every value of the independent variable, x, corresponds with one value of the function, f. We symbolize the derivative of this function as f'(x). Sometimes a textbook will say, "y = f(x)," which indicates that the variable y is a function of the variable x. In that case, we don't care so much about how x and y might vary in time, but rather how y varies as x varies. In other words, we take time out of the equation and focus on how x and y covary. Consider the example where y = f(x) = |x|.

You might recognize the equation, y = |x| as one for the absolute value function, which does not have a derivative at x = 0. Looking at the graph in Figure 11.4, we can see why. Namely, there is no tangent line at x = 0. On the left side of the graph, the slope is negative; on the right side of the graph, the slope is positive. Thus, the instantaneous rate of change (as represented by the slope of the tangent line) is indeterminant when x = 0. Let's see what happens when we apply Newton's method to this function.

The equation describes the way two variables covary, and we want to compute $\dot{y}/\dot{x}$, which describes the rate at which y varies relative to the rate at which x varies. At any moment in time, x varies to become $x + \dot{x}$o and y varies to become $y + \dot{y}$o. Because the two variables co-vary according to the equation y = |x|, the two new values, $x + \dot{x}$o and $y + \dot{y}$o, also satisfy the equation; that is, $y + \dot{y}$o = |$x + \dot{x}$o|. Because the absolute value makes both negative and positive values of $x + \dot{x}$o positive, we have two possibilities: $y + \dot{y}$o = ($x + \dot{x}$o), or $y + \dot{y}$o = −($x + \dot{x}$o). In both equations, we can subtract out the y and x terms because, in the first equation, y = x, and in the second equation, y = −x. We are left with $\dot{y}$o = $\dot{x}$o, or $\dot{y}$o = −$\dot{x}$o. Thus, after dividing out o, we have $\dot{y}/\dot{x}$ = 1 or $\dot{y}/\dot{x}$ = −1.

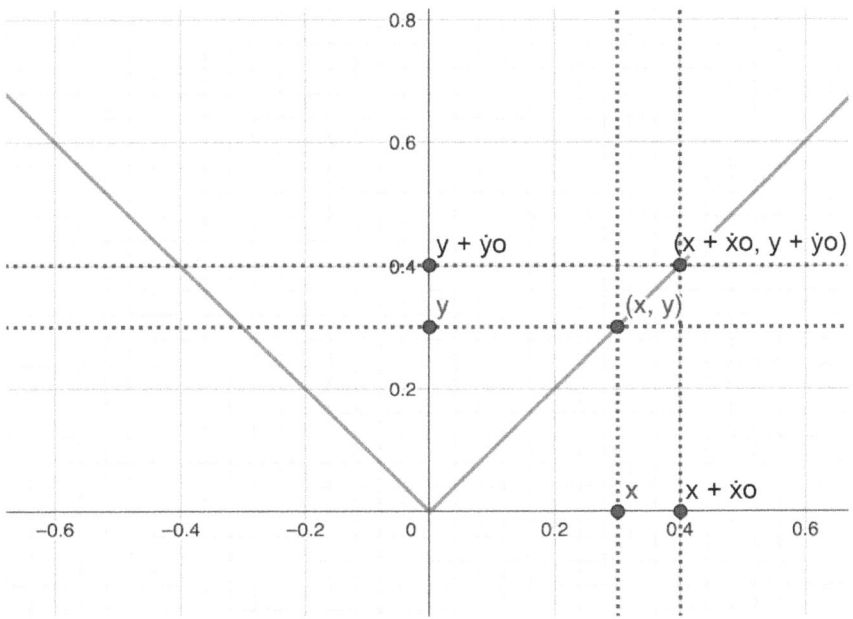

Figure 11.4 Graph of absolute value function and its instantaneous rate of change.

Notice that, in computing the derivative, we never used $y = f(x)$. Using Newton's method, we don't need the variable y to be a function of the variable x in order to find $\dot{y}/\dot{x}$. We can take derivatives of relations that are not functions, such as the relation $x^2 + y^2 = 1$, which defines the unit circle in the Cartesian plane. We consider that example next.

Because the two variables covary according to the equation $x^2 + y^2 = 1$, we know that $(x + \dot{x}o)^2 + (y + \dot{y}o)^2 = 1$. If we proceed as Newton would have done, we would get $x^2 + 2x\dot{x}o + (\dot{x}o)^2 + y^2 + 2y\dot{y}o + (\dot{y}o)^2 = 1$. Because $x^2 + y^2 = 1$, this new equation simplifies to $2x\dot{x}o + (\dot{x}o)^2 + 2y\dot{y}o + (\dot{y}o)^2 = 0$. Here, again, we reach the point where Newton tried to have it both ways: o as a unit with which he could divide, and o as having no value. However, if we include the criterion that $\dot{x}o$ and $\dot{y}o$ are measurable by o, we can proceed, leaving $2x\dot{x} + 2y\dot{y} = 0$. Therefore, $\dot{y}/\dot{x} = -x/y$.

Reflection

In Chapter 10, we noted that, at any point on the unit circle, the tangent line is perpendicular to the radius. So, the slope of the tangent line is the negative reciprocal of the slope of the radius. How do Figure 11.5 and the equation, $\dot{y}/\dot{x} = -x/y$, show this?

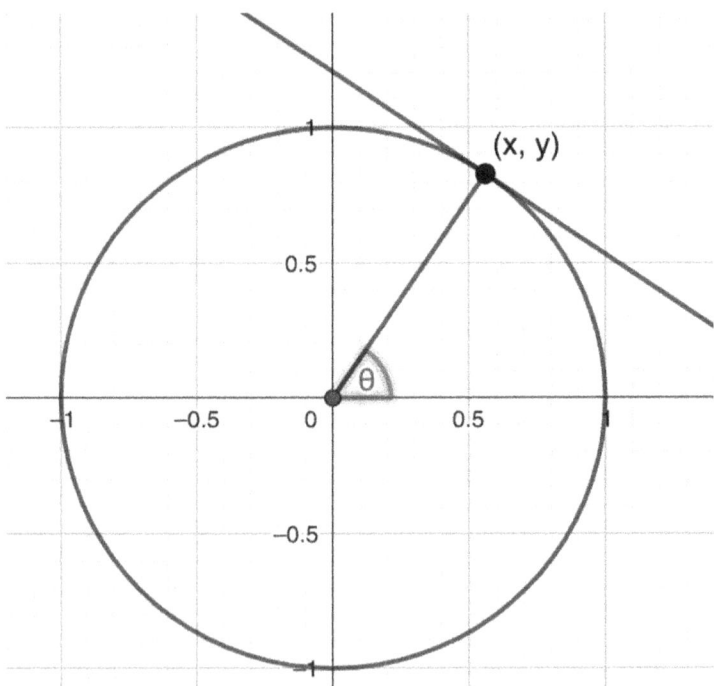

Figure 11.5 Tangent to the circle.

To make a further connection, recall that points on the unit circle can be described by their central angle, θ, so that $(x, y) = (\cos \theta, \sin \theta)$. So, we can represent the same circle, $x^2 + y^2 = 1$, in terms of this angle, using the equation $\sin^2(\theta) + \cos^2(\theta) = 1$. This is just another application of the Pythagorean theorem, $a^2 + b^2 = c^2$, which we derived in Chapter 5: $a = \cos \theta$, $b = \sin \theta$, and $c = 1$ (the radius of the circle). The slope of the radius to any point on the unit circle is given by $\sin \theta / \cos \theta$, which is the definition of $\tan \theta$. The slope of the tangent line, then, is given by the negative reciprocal, $-\cos \theta / \sin \theta$, which is $-\cot \theta$. Thus, for any central angle, θ, $-\cot \theta$ (i.e., $-x/y$) describes how quickly a point on the unit circle rises or falls as it rotates counterclockwise around the circle.

MODERN TRANSLATIONS

We have noted that Newton treated variables as if they varied in time. In modern terms, we describe such variables using parametric equations. We would denote x and y with $x(t)$ and $y(t)$. We would denote $\dot{x}$ and $\dot{y}$ with $\dot{x}(t)$ and $\dot{y}(t)$. When $y = f(x)$, $f'(x)$ would be $\dot{y}(t)/\dot{x}(t)$, which is the modern notation for $\dot{y}/\dot{x}$.

In modern calculus, we rely on limits for a formal definition of $f'(x)$. We choose a kind of dummy variable, h, that represents some finite change in x, and then we consider the following:

$$f'(x) = \lim_{h \to 0} \frac{f(x + h) - f(x)}{h}$$

This equation only makes sense when f is a continuous function (at least at the point x). In other words, we need small changes in h to generate small changes in f, which Newton recognized within his own method. Otherwise, in the limit, we would be attempting to measure a relatively large quantity, $f(x + h) - f(x)$, with an arbitrarily small unit, h—the very thing Newton made sure to avoid.

In modern calculus, formal definitions of derivatives and continuity require a formal definition of limits, and all three of these definitions rely on that same principle that Newton implicitly relied on. For example, we say that $f(x)$ is continuous at x if we can guarantee arbitrarily small changes in f by choosing small enough changes in x. We codify "arbitrarily small" and "small enough," not with infinitesimals but with a kind of game: the epsilon–delta game. We show that, if we are required to keep changes in f to be less than any given value of epsilon (where epsilon can be any real number that is greater than 0), we can do so by keeping changes in x less than some other positive real number of our choosing, which we can delta.

Research on college students' learning of limits indicates that this epsilon–delta game formalizes the reversibility inherent within mathematics in general.[9] Specifically, continuity requires that small changes in x (symbolized by delta) induce small changes in $f(x)$ (symbolized by epsilon). So the formal action progresses in the same direction as the function, domain to range, or input to output. However, students must build their argument for continuity in the opposite direction, defining a new function that takes any given value of epsilon as its input and produces a suitable value of delta. In other words, delta is a function of epsilon. Thus, we find once again that reversibility is a defining characteristic of mathematics and the coordinations of mental actions that constitute it.

As we generalize calculus to higher dimensions, we begin to understand derivatives as linear functions that approximate nonlinear functions. Thus, linear algebra (discussed in Chapter 9) and calculus (discussed in this chapter) become closely aligned. Looking back at Figures 11.2 and 11.5, recall that we represented derivatives as slopes of tangent lines, where a line in the plane represents a linear transformation from the real line to another copy of the real line, and the slope of that line describes the transformation. Thus, we can think about the derivative as a 1x1 matrix, which dilates the real line by a factor of f'(x). In higher dimensions, for functions that map an m-dimensional space to an n-dimensional space, we could represent the derivative as an m × n matrix. Within that framework, opaque notations, such as dx and dy, become clearer, as they designate vectors, not infinitesimal quantities.[10]

THE FUNDAMENTAL THEOREM OF CALCULUS

Most of the notation we learn in calculus classes comes from the 17th-century German mathematician, Leibniz, who along with Newton is credited with inventing calculus. In this modern notation, we can state its fundamental theorem as follows:

$$\int_a^b f'(t)dt = f(b) - f(a)$$

This definition assumes that we can find real values of f'(x) over some interval of the real line, from a to b, which implies that f(x) is continuous within that interval.

Calculus students often interpret the fundamental theorem as a simple statement that the integral (symbolized with the elongated s, for summation; the German word for *sum* is *summe*) is the antiderivative. In other words, the integral and the derivative undo one another, like inverse operations, leaving only the original function, f, evaluated at the boundary points a and b. This conception is not wrong, but it overlooks the full power of the fundamental theorem of calculus and why we might consider it fundamental. To understand the power of the theorem, we return to mental actions that define derivatives and integrals.

Consider the situation of a car, speeding up at a constant rate, from 0 miles per minute to 1 mile per minute, over a period of 1 minute. That is, the car is stopped at the beginning of that minute, and at the end of that minute, it is traveling at 60 miles per hour (1 mile per minute). How far has the car traveled at the end of that minute?[11] The answer is almost immediately given by the fundamental theorem of calculus if we understand all that it signifies.

In this case, the boundary points, a and b, are 0 and 1, respectively. $f'(t) = t$ because speed (in miles per minute) changes at the same rate that time does (in minutes). So, if we can derive the function, $f(t)$, that has the function $f'(t) = t$ as its derivative, all we would need to do is evaluate that function at 1 and 0, and take the difference, to find how far the car has traveled (in miles). We've saw a similar function in Chapter 9: the hourglass problem illustrated in Figure 9.5. Both functions have rates of change that change at a constant rate of change. These are quadratic functions, and in this case, $f(t) = t^2/2$. Thus, $f(1) = 1/2$, $f(0) = 0$, and so the car travels half a mile in that minute.

To understand these relations, consider again the derivative as an instantaneous rate of change. We develop an understanding of instantaneous rate of change through covariation, as discussed in Chapter 9. The fundamental theorem of calculus describes an accumulation of these instantaneous rates of change through a kind of summation.

The graph shown in Figure 11.6 is not a graph of distance versus time but rather speed versus time. It represents how the car begins with a speed of 0 miles per minute, when $t = 0$, and then accelerates smoothly to a speed of 1 mile per minute when $t = 1$. It represents the rate of change in distance at each moment in time. How do we translate these rates of change in distance into actual distance traveled?

One way to reason through the situation is to average the staring speed and the ending speed, thus determining that the average speed over the full minute is half a mile per minute. So, in one minute, the car would travel half a mile. This kind of reasoning works when the rate of change is linear, as it is in this case (the graph is

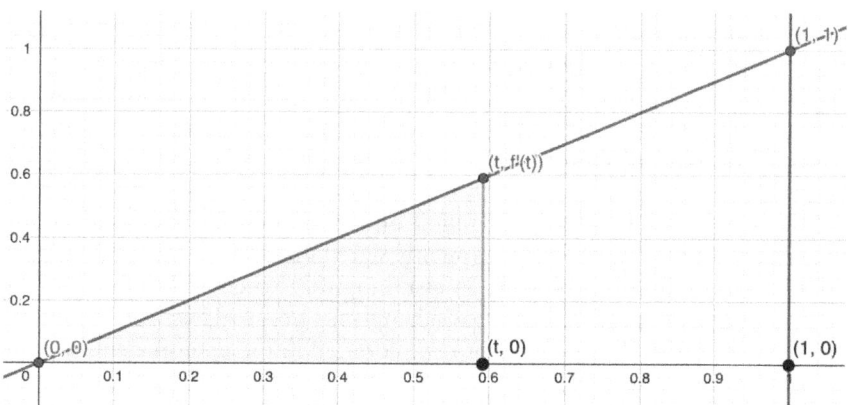

Figure 11.6 Accumulating rates of change.

a straight line), but it doesn't work in general because rates of change don't always change at the same rate. The average of the starting and ending values might not give the average over the entire length of time. To generalize this way of reasoning, we need to account for the way the car accrues rates of change over time, and this is exactly what the integral does.

As noted earlier, the integral involves a kind of summation. It sums up values of a function (in this case, the function $f'(t) = t$) over time (in this case) even as those values vary in time. In terms of mental actions, it involves "coordinating the accumulation of a function's input variable with the accumulation of instantaneous rate-of-change of the function from some fixed starting value to some specified value."[12] Figure 11.6 represents this accumulation through an accrual of area, beginning from $t = 0$. By the time $t = 1$, the integral will have accumulated an area of 1/2 miles, as represented by the area of the larger triangle.

A complete account of mental actions undergirding the fundamental theorem of calculus would require us to consider accruals at finer and finer grain sizes, as we did with rates of change. It would also require us to explicitly account for all the units involved in coordinating accumulations with rates of change. Some of this work has been conducted by the researchers cited here. Some of this work is ongoing. Along the way, we are accruing a complete account for mathematics, from counting to calculus, as a product of our own mental actions—yours and mine.

SUMMARY

It's tempting for us, as teachers and as humans, to project our structures onto others—to suppose, for example, that children come complete with a real number line. To make such an assumption, that real numbers are somehow real, is to forget the labors of ancient Greek mathematicians in constructing numbers on the continuum. We might consider the continuum to be innate because it requires only a sweep of attention, but then we don't need to define it as an uncountable collection of points. What we do need is the possibility of recursively partitioning lengths on the continuum into finer and finer units.[13] Without infinitesimals, we need some kind of limiting processes to describe continuity, derivatives, and integrals.

Newton's approach to calculus relied on infinitesimals, but it made implicit assumptions that we can now find explicitly stated in modern definitions of limits, continuity, and derivatives. Namely, in assuming that $\dot{x}o$ could be measured by o, Newton's approach implied that small changes in one variable correspond with small changes in the other variable, as now codified in the epsilon–delta definition of limits and continuity.

All these calculus concepts—limits, continuity, derivatives, and integrals—return us to the kind of mental coordination introduced in Chapter 9: covariation.

When two quantities vary smoothy in time, or at least in relation to one another, we gain new tools for describing ways that they covary and for identifying invariances in that covariation. We can consider the way the instantaneous rates of change in one variable relative to the other. Conversely, we can accumulate these instantaneous rates of change to reproduce values of one variable as a function of the other.

Investigating the fundamental theorem of calculus, reminds us once more that mathematics is dynamic, as well it should be, because it is based in action. Just as graphs emerge from our coordinations of covarying quantities, the fundamental theorem of calculus emerges from a coordination of the mental actions that undergird instantaneous rates of change and those that undergird instantaneous accruals of a quantity. Like addition and subtraction, derivatives and integrals indicate the reversibility of mathematical operations, but their reversibility reaches far deeper than a simple cancellation rule. It pushes units and measurement to their limits, nearing the infinitesimal.

Activities

Activity 1: Take the derivative of $x(t) = 3t$ the old-fashioned way, using the example illustrated in Figure 11.2 as a guide. What does the value of the result, $\dot{x}$, represent in terms of the original equation? What does it say about the invariant relationship between the variables x and t?

Activity 2: A quadratic equation of the form $y = x^2 + c$ has two real roots if and only if c is negative. For example $y = x^2 - 1$ has two real roots, and $y = x^2 + 1$ has no real roots. How does continuity and the IVT guarantee this result?

Activity 3: If we think about the derivative as a 1×1 matrix that provides a linear approximation of a nonlinear function, what should be the derivative of a linear function. For example, what should be the derivative of $y = x$, or $y = -2x$?

Activity 4: Imagine a circle expanding in time, starting from a radius of $r = 0$ to a radius of $r = t$. What would the following expressions represent?[14]

$2\pi t$

$\int_2^3 2\pi t\, dt$

$\int_0^x 2\pi t\, dt$

NOTES

1.   In standard analysis, we define $0.9\overline{9}$ as a limit. Accepting this definition, the definition of limits, and the Archimedean property (that infinitesimals are not real numbers), we have no logical recourse but to accept that $0.9\overline{9} = 1$ (Norton & Baldwin, 2012). Robert Ely (2010) reported on a related study, on a college student's "nonstandard conceptions of the real number line" (p. 117). Ely showed that the student's conceptions, though nonstandard, matched up well with Leibniz's concepts for infinitesimal calculus, which were similar to Newton's.

2.   Algebraic numbers are those that can be generated as roots of polynomial equations, with rational coefficients. For example, the cubed root of 2 is algebraic (although not constructible) because it is a root of $x^3 - 2 = 0$. Nonalgebraic real numbers are called transcendental numbers.

3.   In "The structure of the continuum," Brouwer (1930) framed the debate as follows: "the arithmetic or geometric continuum has from antiquity been regarded as something given, but there has been little understanding and agreement about the microscopic content of this continuum" (in Mancosu, 1998, p. 54).

4.   Hint: consider the distance from each point along the rope to one of the posts, before and after the rope is moved. We can define a function, $f(x)$, that takes the difference of those two distances, at each point along the rope, x. If $x = 0$ is the point along the rope that was attached to the post, and $x = 1$ is the point along the rope that was attached to the other post, we have $f(0) \leq 0$ and $f(1) \geq 0$. By the IVT, $f(c)=0$ for some point c, between 0 and 1.
Interestingly, we can generalize this problem to a rug that covers the entire floor of a rectangular room. If the rug is picked up, crumpled up (but not cut), and thrown into a pile within a room, there will be some point on the carpet whose distance to all four walls is exactly what it was before.

5.   Núñez, Edwards, and Matos (1999, p. 54) refer to this as the intuitive definition of continuity.

6.   In the modern, epsilon–delta definition of continuity, we say that a function, $f(x)$, is continuous at a point, a, if the following criterion is satisfied: for all positive values of epsilon, there is a positive value of delta, such that the distance between $f(x)$ and $f(a)$ is less than epsilon whenever the distance between x and a is less than delta.

7.   In 1817, the Bohemian mathematician Bernard Bolzano was the first mathematician to formally prove the IVT. In so doing, he argued that we cannot take IVT as a definition of continuity: "The continuous function is one which never reaches a higher value without first going through all lower values. . . . That is certainly a very true assertion, but it cannot be viewed as a definition the concept of continuity: rather, it is a theorem about continuity" (in Russ, 1980, p. 162).

8.   The real issue here is that o is not a real number at all because it violates Archimedean property. This property states that, for any real number, r, there is an integer, N, such that $r \times N > 1$. Infinitesimal and infinite numbers are allowed within the hyper-real system, but this system was not invented until the 1960s (see Robinson, 2016) and few mathematicians use it.

9.   See, for example, Roh (2010).

10.   Munkres (1991) provides a thorough review of differential forms, tensors, and their connections to linear algebra, in *Analysis on Manifolds*.

11.   This task was adapted from a similar one in Thompson, 1994. That research article also details a pedagogical approach to teaching the fundamental theorem of calculus.

12. This quote comes from Carlson, Smith, and Persson (2003, p. 167), who describe a hierarchy of mental actions that generate a complete understanding of the fundamental theorem of calculus.

13. Shin, Lee, and Steffe (2020) elaborate on Les Steffe's original conception of recursive partitioning and its role in problem-solving.

14. This task was adapted from a similar one in Carlson, Smith, and Persson (2003).

REFERENCES

Brouwer, L. E. J. (1930). *Die Struktur des Kontinuums.* Vienna.

Carlson, M. P., Smith, N., & Persson, J. (2003). Developing and connecting Calculus students' notions of rate-of change and accumulation: The fundamental theorem of Calculus. *International Group for the Psychology of Mathematics Education, 2*, 165–172.

Ely, R. (2010). Nonstandard student conceptions about infinitesimals. *Journal for Research in Mathematics Education, 41*(2), 117–146.

Mancosu, P. (1998). *From Brouwer to Hilbert: The debate on the foundations of mathematics in the 1920s.* New York: Oxford University Press.

Munkres, J. R. (1991). *Analysis on manifolds.* Cambridge, MA: Westview Press.

Norton, A., & Baldwin, M. (2012). Does 0.999 . . . really equal 1? *The Mathematics Educator, 21*(2), 58–67.

Núñez, R. E., Edwards, L. D., & Matos, J. F. (1999). Embodied cognition as grounding for situatedness and context in mathematics education. *Educational Studies in Mathematics, 39*(1), 45–65.

Robinson, A. (2016). *Non-standard analysis.* Princeton, NJ: Princeton University Press.

Roh, K. H. (2010). An empirical study of students' understanding of a logical structure in the definition of limit via the ε-strip activity. *Educational Studies in Mathematics, 73*(3), 263–279.

Russ, S. B. (1980). A translation of Bolzano's paper on the intermediate value theorem. *Historia Mathematica, 7*(2), 156–185.

Shin, J., Lee, S. J., & Steffe, L. P. (2020). Problem solving activities of two middle school students with distinct levels of units coordination. *Journal of Mathematical Behavior, 59*.

Thompson, P. W. (1994). Images of rate and operational understanding of the fundamental theorem of calculus. *Educational Studies in Mathematics, 26*(2), 229–274.

12

The Wonderful Gift of Mathematics

In 1859, scientists discovered the planet Vulcan. Because this small, volcanic planet orbited the Sun so closely (inside of Mercury's orbit), they could only catch glimpses of it, but scientists knew it must be there. They knew it was there because Mercury's orbit had a wobble in it that violated the strict elliptical path demanded by Newtonian physics.[1]

In orbiting the Sun, planets obey the inverse square law of gravity: the force of gravity between two massive bodies—such as Mercury and the Sun—is inversely proportional to the square of the distance between them $(1/r^2)$. Galileo formulated this law, and Newton extended into the heavens. From the inverse square law, Newton mathematically derived that planets must move in elliptical orbits. However, Mercury doesn't obey the law, and there's only one Newtonian explanation: there must be some third massive body (enter Vulcan) exerting its own gravitational force on Mercury.

We can laugh now, but at the time, many scientists took Vulcan very seriously, especially because this was exactly how they had discovered Neptune. In a sense, Neptune was invented before it was discovered. Before 1846, there were seven known planets: Mercury, Venus, Earth, Mars, Jupiter, Saturn, and Uranus. When scientists noticed that Uranus's orbit wobbled, they invented Neptune as an explanation. Trusting mathematics, Newtonian physics, and the inverse square law, they

DOI: 10.4324/9781003181729-13

precisely located the new planet from the observable effect it had on Uranus' orbit. They looked and, lo and behold, Neptune was there![2]

Scientists throughout history have marveled at this sort of prediction: "the unreasonable effectiveness of mathematics in the natural sciences."[3] Mathematics proved perfectly reliable when Newton used it to explain the elliptical orbits of planets, based on the inverse square law. Even more uncanny, mathematical computations had predicted the precise location of Neptune, When things went wrong, in the case of planet Vulcan, mathematics stood blameless. Rather, it was Newton's physics that needed replacing.

Mercury's irregular orbit was not the influence of some unseen planet (i.e., the fictitious planet Vulcan). Instead, it provides evidence for Einstein's theory of relativity in which massive objects, like the Sun, warp space-time. Because Mercury orbits the Sun so closely, we can actually see the effect of the Sun's gravity in warping its path. Observations like these led scientists to replace Newton's physics—however useful it might have been up to that point in time—with Einstein's relativity. Physics was revised, but mathematics remained intact. As math historian Morris Kline put it, "there is one quality that distinguishes most of mathematics from physical theories. Whereas in science there have been radical changes in theories, in mathematics most of the logic, number theory, and classic analysis has functioned for centuries."[4]

What justification do we have for trusting mathematics more than scientific theories? How can we explain mathematics' uncanny ability to predict phenomena in the natural sciences, such as physics and chemistry? We address these questions in the next two sections while highlighting themes that have arisen from prior chapters. In all, we highlight seven themes, in seven sections, summarizing a journey of mathematical empowerment that began with seven yellow bricks.

PROJECTING OUR STRUCTURES INTO THE WORLD

Albert Einstein put the question like this: "How can it be that mathematics, a product of human thought, independent of experience, is so admirably adapted to the objects of reality?"[5] Eugene Wigner provided the Platonist response: "The miracle of the appropriateness of the language of mathematics for the formulation of the laws of physics is a wonderful gift which we neither understand nor deserve."[6] Not to be ungrateful, but human knowledge always proceeds by looking a gift horse in the mouth.

In describing mathematics as a language or a gift, Galileo, Wigner, and others avoid taking the question seriously. Like most physicists and mathematicians, they would rather focus on doing mathematics than on questioning its foundations:

> The typical "working mathematician" is a Platonist on weekdays and a
> formalist on Sundays. That is, when he is doing mathematics, he is convinced
> that he is dealing with an objective reality whose properties he is attempting to

determine. But then, when challenged to give a philosophical account of this reality, he finds it easiest to pretend that he does not believe in it after all.[7]

Platonism serves as an unsupportable but motivating fantasy for many mathematicians. Einstein appears to be an exception, admitting mathematics as a product of human thought. Maybe this should not surprise us since, in developing his theory of relativity, Einstein brought us, as human observers, into the equation.

Einstein challenged our conceptions of time and space around the same time Piaget and Inhelder were studying how children construct them (see Chapter 2). Collectively, their work demonstrated that our own conceptions of time and space provide frames of reference for understanding what we see. Once we have constructed these fundamental structures, we project them into the world by using them to assimilate our experiences in the world.

We have discussed three kinds of projection in this book. In Chapter 3, we introduced projections that emanate from a point, as used in projective geometry. In Chapter 5, we considered parallel projections that destroy dimensions and provide an inverse for sweeps. These first two kinds of projection are mental actions that we can coordinate with other mental actions to construct mathematical structures, including space. The third kind of projection, discussed in the Introduction and here, describes the way we project those structures into the world. To illustrate, let's revisit an example from the Introduction.

Table 12.1 shows the first eight rows of Pingala's triangle. In the Introduction, we asked, "What do you see?" Now, we ask, "What do you see now?" Maybe before, you saw the symmetry of numbers across each row but didn't make a connection to powers of 11 (e.g., the fourth, highlighted row shows 11 cubed). Maybe now, having engaged in activities suggested in Chapter 6, you see the coefficients of binomial expansion (e.g., $(m + n)^3 = 1m^3 + 3m^2n + 3mn^2 + 1n^3$). Maybe you also see triangular numbers in the third column and a connection to Fermat primes (as discussed in Chapter 10) in the bottom row.

Table 12.1 Revisiting Pingala's triangle

1							
1	1						
1	2	1					
1	3	3	1				
1	4	6	4	1			
1	5	10	10	5	1		
1	6	15	20	15	6	1	
1	7	21	35	35	21	7	1

If you look back at figures across the chapters (e.g., areas of squares in Chapter 5, intersecting chords in Chapter 7, or graphs of functions in Chapter 11) and see something new, it's not because they have changed but because you have changed. That is, you have constructed new structures for assimilating them. In that sense, you can project your mathematical structures into the figures and anything else that you see in the world. In this way, the problem of explaining the unreasonable effectiveness of mathematics in the natural sciences is turned on its head. Mathematics is not a language written into the fabric of the universe; it is constructed by us and used to make sense of our experiences in an otherwise formless world.

Returning to Neptune, scientists quite literally projected it out into the heavens as a mathematical structure. They derived its existence through mathematical formulas and then aimed their telescopes precisely where Neptune should be, according to their calculations. However, if that same mathematics were written into the fabric of the universe, they also would have found Vulcan where it belonged. It wasn't there because their constructions of space and time didn't quite fit outer space (especially a warped four-dimensional space-time), and why should they have?

We construct space and time based on our own activities on Earth. It is reasonable for us to extend those structures beyond our experience, in an attempt to model and explain the unknown (e.g., planetary motion). However, it seems unreasonable to expect those structures, and mathematical models based on them, to fit the universe perfectly. Instead, we should be amazed that they work at all. Piaget explained this uncanny power of mathematics as follows:

> Pure logic and pure mathematics are forever capable of transcending experience. . . . But as human action is that of an organism which is part of the physical universe, we understand also why these unlimited operatory combinations so often anticipate the experience, and why when they encounter each other there is harmony between the characteristics of the object and the operations of the subject.[8]

Specifically, Piaget was answering Einstein's question. If mathematics is a product of our own mental activity, why is it so useful in building scientific models of the world outside of us? Piaget's answer is that, even though we construct mathematics on the basis of our own mental actions, those actions initially arise through our physical actions in the world. Moreover, when we internalize and coordinate those sensorimotor activities as mental actions, we organize them within structures that have two vital characteristics: reversibility and composability.

REVERSING AND COMPOSING ACTIONS

When a scientific prediction goes awry (as it did in the case of Vulcan), we don't have to question the underlying scientific theory. We could instead toss out the mathematical equations and calculations that led from the scientific theory to the prediction. We never do this and for good reason. Mathematics is reliable in a way that scientific theories are not.

In science, reliability is repeatability. It means that someone else could conduct a similar experiment, under similar conditions, and expect similar results. This reliability is never perfect because the conditions of any two experiments are never precisely the same. However, in mathematics, every experiment is perfectly repeatable and thus perfectly reliable, thanks to the reversibility of mathematical actions. The reversibility of mathematical actions guarantees that we can always return to a starting point and repeat the same result, even at the end of a long chain of compositions. For example, consider fraction multiplication. In Chapter 4, we described the product $2/3 \times 3/2$ as the composition of four reversible mental actions: $P_3 I_2 P_2 I_3$. Not only are the two fractions multiplicative inverses of each other, but because we can reverse each of the mental actions that comprise their product, we can move forward and backward through each one of them.

If reversibility endows mathematics with perfect reliability, composability provides it with the potential for reliable extension. Consider the geometric example of reflection from Chapter 2. By composing mental actions of reflection with one another, we generated every isometry of the plane. Then we considered what might happen if we composed two rotations with different centers of rotation. We didn't even have to perform the rotations to determine that their composition would generate another rotation, about some third center. We could do this because we knew each rotation was composed of two reflections. By composing a chain of four reflections, we extended them to predict the result of a mental action we had not yet performed. This kind of extension is not so different from Newton's extension of Galileo's inverse square law to the planets, except our mathematical extensions are perfectly reliable.

Neither mathematics nor science is true, in the sense that neither of them perfectly fits the universe. Mathematics is not the language of the universe. If it were, we would have only one kind of geometry and not the several geometries that Klein classified through his Erlangen program (see Chapter 3). However, mathematics enables us to extend our actions beyond immediate experience and anticipate the effects of long chains of actions. When their application goes awry, it is their application—and not the chain of actions—that is at fault. We can abandon Platonism and still recognize this unique power of mathematics:

> In all, the chief apparent advantage of Platonism, which is to account for the objective robustness of logico-mathematical entities and structures, is guaranteed in the same way by the concept of the general and internal co-ordinations of actions and operations. The hypothesis that ideal entities are external is thus unnecessary to guarantee the independence of structures in view of the free will of individual subjects.[9]

THE UNITY OF SPACE AND NUMBER

Dictionaries typically define *mathematics* as a collection of related studies, such as algebra and geometry, or the study of a collection of related objects, such as shapes and numbers.[10] These definitions emphasize the plurality of the word *mathematics* but fail to establish how those studies and objects are related under a single umbrella called math.

A better definition comes from mathematician Lynn Arthur Steen, who defined *mathematics* as the study of patterns.[11] We see patterns in both shape and number. We see patterns everywhere, but where do these patterns come from? When we see a spider web, we might admire the spider as a geometer. We might see lines of symmetry and lines that spiral out into larger and larger polygons. Does the spider see them too? Was this the spider's intent?

My brother is a scuba diver. He once told me he could find sunken ships and other human creations underwater by looking for straight lines. Nothing in nature (since Einstein, not even light) moves in a straight line, and yet much of mathematics is built on straight lines. Since the Greeks, we have placed numbers on a number line, and all Euclidean geometry is based on intersecting circles and lines. Entire branches of mathematics—like linear algebra (discussed in Chapter 8) and calculus (discussed in Chapter 11)—involve linear approximations of nonlinear functions, so that we can better understand those functions (e.g., their rates of change). Mathematics is a human creation.

We use lines, circles, and units to create patterns in the world. The beauty of the spider web comes from the joy we experience in creating the patterns we see in them. After all, art always invites you in.[12] If there is a distinction between art and nature maybe it is the intention of the artist, but your intention, as observer, is ever-present, and the line between observing and creating is now blurred.

Along lines, we can move forward and backward, in a perfectly reliable way. We can transform lines in space and transform them back. With units too, we can move forward and backward, by continuing their iterations or disembedding some number of them. We can transform units into other units and transform them back again. With both shape and number we can generate patterns by composing these objects and their transformations. So, we might define *mathematics* as the study of patterns, but we might just as well define *mathematics* by the kinds of structures we construct in order to create and see those patterns. Mathematics is a unified body of knowledge, across space and number, because we construct its objects of study on the basis of reversible and composable mental actions.

The unity of space and number is beautifully demonstrated in the complex plane, introduced in Chapter 8. Therein, numbers achieve a geometry complete with reflections, rotations, dilations, and translations. We saw in Chapter 10, that we could represent points on the unit circle as vectors in the complex plane, with $\cos\theta + i\sin\theta$, where θ represents the direction of the vector. All the complex numbers on the unit circle have a magnitude of 1, but we can dilate the circle, by a factor of r, to describe the coordinates for any point in the complex plane:

r(cosθ + isinθ). Thus, a complex number is a single number that describes a point in the plane by its direction and magnitude—its angle, θ, and its radius, r.

We saw in Chapter 10 that squaring a complex number on the unit circle doubles its angle. Now, we see that squaring a complex number, in general, doubles its angle and squares its radius: if $z = r(\cos\theta + i\sin\theta)$, then $z^2 = r^2(\cos2\theta + i\sin2\theta)$. Even more generally, if we multiply any two complex numbers, we just add their angles and multiply their radii.

We created the geometry of the complex plane by representing i as a 90-degree rotation of 1, about the origin, 0 (see Chapter 8). We can rely on this geometry to make elegant connections between number and shape. For example, consider the fourth roots of 1. These roots include 1 and −1 because $1^4 = (-1)^4 = 1$, but since the invention of the complex plane, they also include i and −i, for the same reason: $i^4 = (-i)^4 = 1$. In general, we can generate n nth roots of 1, and they will form the vertices of a regular n-gon on the unit circle.[13] As noted in Chapter 10, we can construct some of these regular n-gons using Plato's rules for construction and others we cannot, but we can describe all of them through the simple expression $\sqrt[n]{1}$ (for the example of n = 5, see Figures 10.4 and 10.6 in Chapter 10).

Neuroscience, too, suggests the unity of space and number. Both numerical and spatial manipulation (e.g., mentally rotating a cube) closely correlate with heightened neural activity in the parietal lobe, particularly in and around the intraparietal sulcus (see Figure 12.1).[14] The parietal lobe sits just behind the sensorimotor cortex, which is responsible for controlling all bodily movement. The intraparietal sulcus aligns with the region of the sensorimotor cortex that controls hand movement. Not coincidentally, the intraparietal sulcus also plays key roles in hand–eye coordination and tool use. This correspondence supports the hypothesis that human mathematical ability evolved from tool use,[15] and it emphasizes the importance of manipulatives in supporting children's mathematical development.

The close correspondence between numerical and spatial reasoning has led some researchers to posit an innate mental number line: "a spatial representation of number magnitude along an analog number line which is assumed to be activated automatically whenever we encounter a number."[16] A few researchers go so far as to suggest that this mental number line comes ready-made with all the real numbers.[17] However, given the historical development of real numbers reviewed in Chapter 11, it is more plausible that humans are born with a numberless continuum on which they construct numbers through their own activity. More plausible still, children construct the continuum and space too, during infancy, through their own movement, as described in Chapter 2.

PRESERVING AND TRANSFORMING UNITS

We would have no numbers without units. Starting from a unit of 1, we produce every whole number through iteration. We produce fractions by partitioning that

Frontal Lobe   **Parietal Lobe**

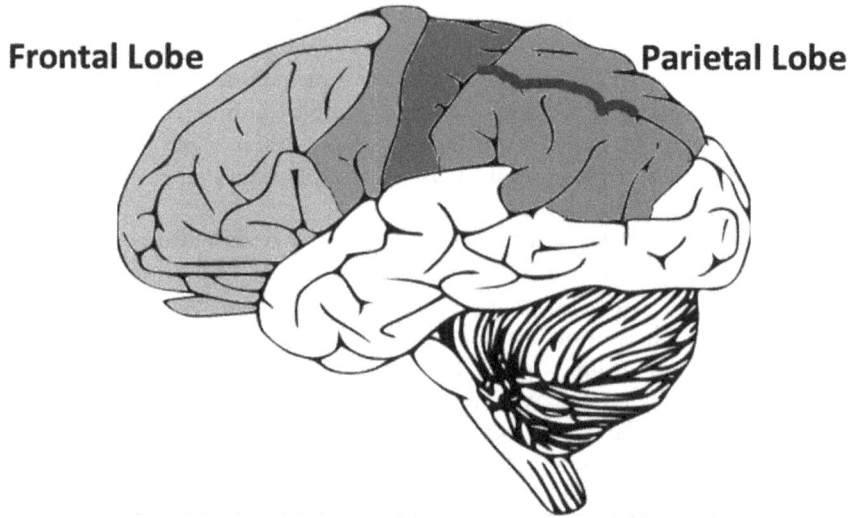

Figure 12.1 Frontal and parietal lobes of the brain.

unit of 1 into various unit fractions and iterating those units. We produce integers and complex numbers by iterating new units, such as i, in new directions. When working with these units and the numbers they produce, we rely on two principal binary operations: addition and multiplication. In reasoning additively, we preserve units; in reasoning multiplicatively, we transform them.

In Chapter 1, we described addition as the continuation of a count via iteration. $2 + 2 = 4$ because, starting from two iterations of 1, iterating two more 1s yields $1 + 1 + 1 + 1$, or four iterations of 1, which is how we construct 4 in the first place. When learning to add fractions in school, we are taught to find a common denominator because, in so doing, we produce a common unit for measuring each fraction in the sum. For example, in determining the sum $2/3 + 1/2$, we seek a unit fraction for measuring both 2/3 and 1/2; because 2/3 is four iterations of the unit 1/6 and 1/2 is three iterations of the unit 1/6, we know that $2/3 + 1/2$ is five iterations of the unit 1/6. Finally, when adding vectors, we add them component-wise because each component has its own unit (see Chapter 8). In each case, the units are preserved, through iteration.

In multiplication, the situation changes. As described in Chapters 1 and 4, whole number multiplication and fractions multiplication transform units of measure. For example. We can measure 6 in six units of 1 or three units of 2, and the product 3×2 represents the transformation from units of 2 to units of 1. When we include negative numbers and complex numbers, we have units in different directions, so the transformation of units can change both magnitude and direction. We can generalize this idea to n–dimensional vectors and linear algebra, wherein matrix multiplication describes the transformation of vectors, along with their n units.

Formally speaking, the principal operations of addition and multiplication are brought together within an algebraic structure called a field (see Chapter 10). Therein, the only way to relate addition and multiplication is through the distributive property: $a \times (b + c) = a \times b + a \times c$. However, in studying the additive and multiplicative reasoning of children, Confrey and Smith described the situation differently—as two different worlds.[18]

Children build their additive worlds by iterating units of 1. Children build their multiplicative worlds by recursively transforming units, which Confrey refers to as "splitting."[19] For example, a child might produce 8 additively, as $1 + 1 + 1 + 1 + 1 + 1 + 1 + 1$, or they might produce it multiplicatively through recursive splitting. In this case, each split would involve transforming units of 1 into units of 2: doubling, doubling again, and doubling once more.

Relating additive and multiplicative worlds involves mapping numbers between them, where 3 in the multiplicative world of doubling corresponds to 8 in the additive world of iterating 1s, because $2^3 = 8$ (see Figure 12.2). Conversely, 8 in the additive world corresponds to 3 in the multiplicative world, because the $\log_2 8 = 3$. Thus, exponential functions, like $f(x) = 2^x$, and their inverse functions—logarithmic functions like $f^{-1}(x) = \log_2(x)$—describe mappings between additive and multiplicative worlds.

By doubling, we can produce powers of 2, like 2, 4, 8, and 16. By tripling, we could produce 3, 9, 27, and 81. By combining such splits, we can produce any whole number. That is to say, every number has a unique prime factorization,[20] which describes how we might build it multiplicatively, as a sequence of transformations of units. For example, $24 = 2^3 \times 3$. We can build 24 multiplicatively by doubling three times and tripling once. Just as iterated 1s are the basic building blocks for the additive world, prime numbers are the basic building blocks for the multiplicative world.

Now that we have some idea about how to map between worlds, we might gain some insight into one of the most surprising theorems about prime numbers. The prime number theorem states that the number of prime numbers between 1 and n is approximately $n/\log(n)$.[21] This result is surprising because it specifies the relative distribution of prime numbers, even though there is no discernable pattern for generating prime numbers.[22] For example, there are four prime numbers less than or equal to 10: 2, 3, 5, and 7. These primes appear patternless, but the prime number theorem accurately estimates their number: $10/\log(10) = 4.3$, which rounds to 4.

Proofs of the prime number theorem often rely on complex analysis, which is surprising in and of itself, given that prime numbers are all whole numbers. The aforementioned geometry of the complex plane provides a clue as to why it might be useful. Clues about the prime number theorem itself might be found in the mapping between additive and multiplicative worlds. Specifically, if we can describe every number additively, as an iteration of 1s, or multiplicatively, through their prime factorizations, the mapping between additive and multiplicative worlds might indicate the relative number of primes within a range of numbers. So, we shouldn't be so surprised that a logarithmic function appears in the prime number theorem.

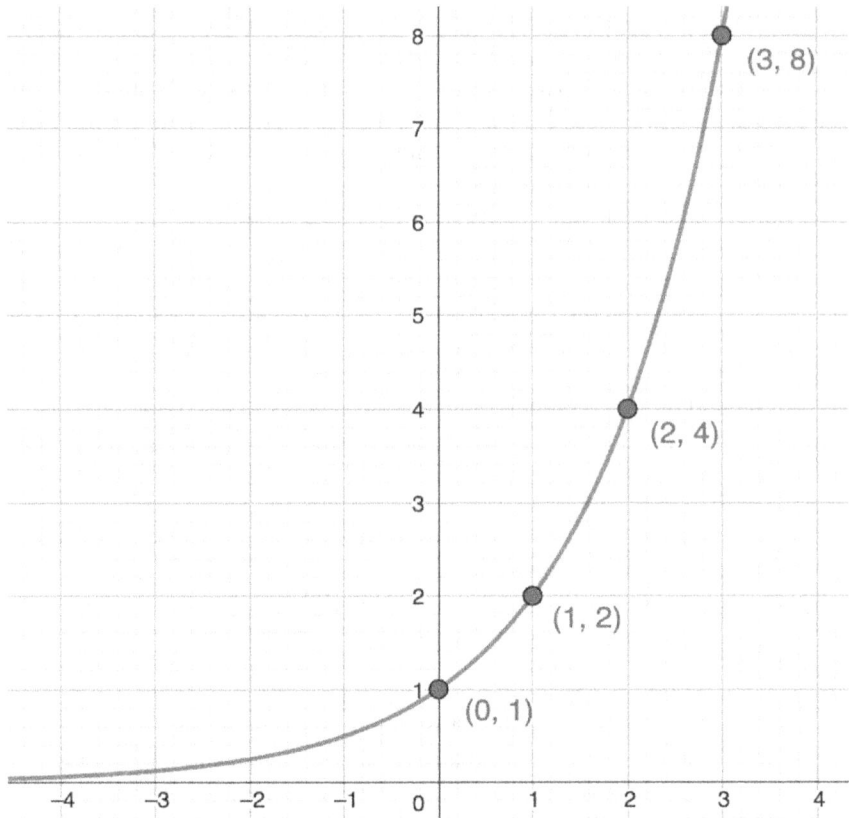

Figure 12.2 Mapping between additive and multiplicative worlds.

THE ALGEBRAIC EXTENSION

One of my math professors in graduate school would stop in the middle of his lectures to recount stories from students past. On one occasion, he conjured a student who claimed to enjoy mathematics, "except for the abstract part." He replied, "math is abstract, unless every time you talk about nine you are talking about nine cows."[23] We begin abstracting number very early in life, so early that we take this abstraction for granted (like we do space). It seems our trouble with abstraction is not abstraction per se, but when we encounter abstract symbols that have no connection to our mental actions.

Fractions and algebra both require a degree of proficiency with units and units coordinations before we can make much meaning of them. As described in Chapter 4, we have to construct new kinds of units, such as unit fractions and unknowns, and operate on them. As described in Chapters 6 and 9, algebraic equations require even more, to coordinate covarying quantities, known as variables. Introducing algebraic manipulation makes sense only when students have learned to operate on unknowns and variables as mathematical objects. Thereafter, manipulating symbols can serve as a proxy for acting on the mathematical objects that they represent.

We call algebra "abstract algebra" once it focuses on the mathematical structures that govern algebraic manipulation. These governing structures, like groups and fields, are abstract in the sense that they are further removed from sensorimotor activity. Whereas learning to count is closely related to such activity (e.g., pointing at items and reciting number words), abstract algebra seems to focus on formal rules (e.g., the distributive property). However, looking back across the chapters in this book, you might find that the structures introduced in abstract algebra closely correspond to your own psychological structures. Groups, in particular, describe the reversibility and composability that characterize mathematical mental actions. When we study abstract algebra, we reflect on psychological structures we have subconsciously used to organize number and geometry from the start.

In performing symbolic manipulations, we sometimes inadvertently produce new combinations of symbols that we had not accounted for. For example, in Chapter 8, we introduced i via the seemingly harmless equation $x^2 + 1 = 0$. Thus, algebra not only symbolizes numbers but can extend them as well. In particular, we can generate the complex plane as a field extension of the real line, by including additive and multiplicative combinations of i. This is precisely what we symbolize in writing complex numbers as a + bi (where a and b are real numbers).

As discussed in Chapter 3, Felix Klein's Erlangen program leveraged abstract algebra to classify geometries as nested groups, further demonstrating the unity of mathematics. But no one contributed more to the development of abstract algebra than Emmy Noether. "She transformed that subject matter, thereby originating a new algebraic tradition—what has come to be known as modern or abstract algebra."[24] One of Noether's protégés, Bartel van der Waerden, captured the unifying power of that tradition in the following quote: "All relations between numbers, functions, and operations become perspicuous, capable of generalization, and truly fruitful, when they are detached from specific examples and traced back to conceptual connections."

HOW MATH BUILDS ON ITSELF

In school, mathematics can seem like a string of courses, each filled with a list of loosely related topics. Sometimes, we get a sense of how one course or topic builds on another (e.g., calculus obviously uses algebra), but this book has attempted to illustrate how math builds on itself psychologically.

Each of the prior chapters has demonstrated how mathematical objects arise through the coordination of your own mental actions. For example, the splitting group (Chapter 4) describes how we construct fractions by coordinating reversible actions of partitioning and iterating, which themselves depend on prior construction of composite units (e.g., in order to partition a whole into five equal parts, we need a composite unit of 5). Conversely, we rely on fractions for constructing real numbers and solving algebraic equations. At any given stage in our development of mathematics, we perform actions on mathematical objects that had been a sequence of actions at a prior stage.

Figure 12.3 offers a simple illustration of the process. The first arrow in the figure represents the way we act on objects we have already constructed (e.g., acting on whole numbers through iterating and partitioning to construct fractions). The second arrow represents the way we coordinate those actions to form new objects. The cycle continues as we perform new actions on those objects. In addition to constructing new objects and thus making meaning of them, psychological benefits for this recursive process extend to the economical use of working memory.

Working memory is a limited cognitive resource that we use to sequence information or actions. Children in middle school can typically hold in mind about five actions and coordinate them sequentially. Adults can typically hold in mind a sequence of seven actions.[25] When we construct new mathematical objects, we organize sequences of action into single chunks so that we don't have to keep in mind each individual action that comprises that object/chunk. For example, when we construct "fractions as numbers in their own right,"[26] we no longer have to imagine a sequence of partitionings and iterations, although we can unpack the fraction into those constituent actions when needed.

Neuroscience studies provide further evidence of the benefits of mathematical construction. These studies show a frontal-to-parietal shift in neural activity as students progress in age and expertise (see Figure 12.1).[27] Neural activity in frontal lobe correlates with working memory: effortful activity in planning out a sequence of actions. Neural activity in the parietal lobe correlates with spatial-numerical structures. Thus, the frontal to parietal shift indicates a decreased dependence on working memory as we construct spatial and numerical objects.

EQUITY IN ACTION

Each chapter in this book provides clues about the historical development of mathematical ideas. From these clues, we might infer the kinds of mental actions ancient mathematicians performed to construct the mathematical objects represented in textbooks. You might think of it as mathematical archeology, which is not so different than what teachers do every day, as they infer a student's

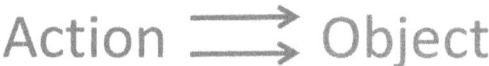

Figure 12.3 Mathematical actions and objects.

reasoning from their solution to a math problem. For example, a closer look at Euclid's *Elements* suggests that he relied heavily on the mental action of shearing to formulate his proof of the Pythagorean theorem (see Chapter 5).

Histories reported in each chapter also indicate cultural influences on mathematical development. For example, in their early invention of linear algebra, ancient Chinese mathematicians certainly benefited from their system of counting rods and use of counting boards, with their arrays of units (see Chapter 8). Furthermore, we find that credit for mathematical developments went to mathematicians in economically powerful countries—often inappropriately, as was the case with linear algebra and "Gaussian elimination."

Likewise in schools, classroom culture affords or constrains opportunities for students to develop mathematically, and teachers must be mindful of the personal cultures their students bring into the classroom. In establishing safe and productive classroom cultures, teachers can also invite students to leverage their personal histories to afford learning opportunities. This book has emphasized that those opportunities consist of opportunities for action.

In Chapters 5 and 10, we considered mathematical developments that depended on Plato's rules for construction. Euclid used these rules as the first three axioms from which he proved the Pythagorean theorem; those rules also determine which numbers and shapes we deem constructible (e.g., constructible regular polygons). Recall that Plato's rules prescribe ways that we can construct lengths and geometric figures from lines, circles, and their intersections. Different cultures might lay down different rules and thus produce very different geometric results. However, the rules and their combinations are not arbitrary.

Underneath every axiomatic system we find prescriptions for coordinating actions. Groups constitute one of the simplest axiomatic systems, and they underlie many of the other systems, such as fields and Euclid's geometric transformations. The group axioms (described in Chapter 3) are fundamental in mathematics because they describe the reversibility and composability of actions. This foundation appears across cultures and within every branch of mathematics. This ubiquity of action in mathematics should go hand in hand with equity in mathematics education.

Equity in mathematics education refers to students' opportunities to access mathematics in personally meaningful ways. Reframing mathematics as a product of our own psychology has direct consequences for mathematics education because it invites students to own mathematics, not merely access it. Mathematics does not come from a textbook or the fabric of the universe, as delivered by renowned mathematicians; it comes from the students' own mental actions. They need opportunities to coordinate those actions.

School math often appears divorced from our own mathematics. There seems to be a disconnect between the activity of creating mathematical objects and students' experiences in classrooms. Over years of schooling, we find a growing disparity between the mathematical objects that students have constructed and the mathematical objects that textbooks represent with words and symbols. For many students, these symbols point to "pseudo-objects" in that they refer to objects the students have not had the opportunity to construct.[28] These students are left with no recourse but to engage in meaningless symbol pushing in order to satisfy behavioral expectations, especially standardized tests.

The history of mathematics teaches us that mathematical constructions typically occur through problem-solving. What problems might have personal meaning for students? What actions might they bring to bear on those problems? And what opportunities, tools, and support do they have for coordinating those actions?

CLOSING

Wigner described mathematics as a wonderful gift. This gift comes to us through affordances to act in the world and to coordinate those actions within reversible and composable structures. It is a gift that keeps on giving, having given us everything from numbers to Neptune. The motivation for writing this book was not simply to marvel at mathematics but also to promote mathematics as personally empowering: personal because we own it, in the same way we own our own actions; powerful because we can extend it into the heavens. None of this makes mathematics easier. After all, coordinating even simple actions, like reflections, can get complicated quickly (see Chapter 2). It only makes mathematics yours.

NOTES

1.   See Darling (2006) for a detailed account of this history, including the history of gravity.

2.   Theoretical physicist Lisa Randall explains numerous and more recent examples like this, in which mathematical models predict the observation of physical phenomena, in her book *Warped Passages* (2006).

3.   This quote is the title of an article by Wigner (1960).

4.   This quote appears in Morris Kline's (1982, p. 333) book, *Mathematics: The Loss of Certainty*.

5.   This quote comes from Einstein's (1921) lecture on "Geometry and Experience."

6.   This is the conclusion Wigner (1960, p. 14) draws at the end of his article (also see Norton, 2015).

7.   This quote comes from mathematician Reuben Hersh (1979, p. 32)

8.   This quote comes from Piaget (1971, p. 72).

9.   This quote comes from Beth and Piaget (1966, p. 294). Evert Beth was a logician who studied the foundations of mathematics.

10.   For example, the Oxford English Dictionary defines *math* as "the science of numbers and shapes."

11.   Steen (1988) elaborated on this idea in an article for *Science* magazine entitled, "the science of patterns."

12.   In her article, "Kant and the Brain," Linda Palmer (2008) describes beauty as follows: "The experience of beauty, then, is the pleasure one takes in one's own mental

activity in apprehending an object, and doing so freely, without the normal constraints of taking it cognitively."

13. See Caglayan (2016) for an investigation of these "roots of unity," using technology.

14. Intraparietal sulcus refers to a segment of the brain that sits in a valley (sulcus) in the middle of (intra) the parietal lobe.

15. Penner-Wilger and Anderson (2013), for example, argue for an evolutionary link between numeral development and manual dexterity.

16. This quote comes from Link, Moeller, Huber, Fischer, and Nuerk (2013, p. 75) as cited in Tabor, Dibley, Hackenberg, and Norton (2020).

17. For example, Gallistel and Gelman (2000, p. 59) express this view by saying that "the non-verbal representatives of number are mental magnitudes (real numbers)." See Ulrich and Norton (2019) for a critique of this perspective.

18. Confrey and Smith (1995) theorize ways that students might build an isomorphism between their additive and multiplicative worlds.

19. Confrey's use of splitting is related to, but differs from, the way we defined splitting in Chapter 5.

20. This result is called the fundamental theorem of arithmetic.

21. Here, we refer to the natural logarithm.

22. Fermat primes (discussed in Chapter 10) were an attempt to systematically generate prime numbers.

23. Thanks to my friend Dr. Wendy Sanchez for recalling this story and to Professor John Hollingsworth for his service to us as his students.

24. These quotes come from Kleiner (2007, p. 92) and van der Waerden (1935, p. 469), as cited by Rowe and Koreuber (2020, p. 84).

25. These estimates come from Pascual-Leone (1970) who described working memory in terms of M-capacity: the number of action schemes a student can hold in mind and coordinate while solving a problem.

26. Recall from Chapter 4, this is how Hackenberg (2007) described the construction of fractions as mathematical objects.

27. For example, Ansari and Dhital (2006) found an age-related frontal-to-parietal shift in neural activity among students, across ages 10 to 20, as they counted collections of dots.

28. Sfard (1991) first used this phrase to distinguish meaningless symbolic manipulation from students' operations on mathematical objects they have constructed.

REFERENCES

Ansari, D., & Dhital, B. (2006). Age-related changes in the activation of the intraparietal sulcus during nonsymbolic magnitude processing: An event-related functional magnetic resonance imaging study. *Journal of Cognitive Neuroscience*, *18*(11), 1820–1828.

Beth, E., & Piaget, J. (1966). *Mathematical epistemology and psychology* (W. Mays, Trans.). New York: Gordon & Breach.

Caglayan, G. (2016). Mathematics teachers' visualization of complex number multiplication and the roots of unity in a dynamic geometry environment. *Computers in the Schools*, *33*(3), 187–209.

Confrey, J., & Smith, E. (1995). Splitting, covariation, and their role in the development of exponential functions. *Journal for Research in Mathematics Education, 26*(1), 66–86.

Darling, D. (2006). *Gravity's arc: The story of gravity from Aristotle to Einstein and beyond.* Hoboken, NJ: John Wiley & Sons.

Einstein, A. (1921). *Geometry and experience.* Lecture for the Prussian Academy of Sciences, on January 27.

Gallistel, C. R., & Gelman, R. (2000). Non-verbal numerical cognition: From reals to integers. *Trends in Cognitive Sciences, 4*(2), 59–65.

Hackenberg, A. J. (2007). Units coordination and the construction of improper fractions: A revision of the splitting hypothesis. *The Journal of Mathematical Behavior, 26*, 27–47.

Hersh, R. (1979). Some proposals for reviving the philosophy of mathematics. *Advances in Mathematics, 31*(1), 31–50.

Kleiner, I. (2007). *A history of abstract algebra.* Boston: Birkhäuser.

Kline, M. (1982). *Mathematics: The loss of certainty.* New York: Oxford University Press.

Link, T., Moeller, K., Huber, S., Fischer, U., & Nuerk, H. C. (2013). Walk the number line–An embodied training of numerical concepts. *Trends in Neuroscience and Education, 2*(2), 74–84.

Norton, A. (2015). The wonderful gift of mathematics. *The Mathematics Educator, 24*(1).

Palmer, L. (2008). Kant and the brain: A new empirical hypothesis. *Review of General Psychology, 12*(2), 105–117.

Pascual-Leone, J. (1970). A mathematical model for the transition rule in Piaget's developmental stages. *Acta Psychologica, 32*, 301–345.

Penner-Wilger, M., & Anderson, M. L. (2013). The relation between finger gnosis and mathematical ability: Why redeployment of neural circuits best explains the finding. *Frontiers in Psychology, 4*.

Piaget, J. (1971). *Psychology and epistemology* (A. Rosin, Trans.). New York: Grossman Publishers (Original work published in 1970).

Randall, L. (2006). *Warped passages: Unravelling the universe's hidden dimensions.* New York: Harper.

Rowe, D. E., & Koreuber, M. (2020). *Proving it her way: Emmy Noether, a life in mathematics.* Cham, Switzerland: Springer.

Sfard, A. (1991). On the dual nature of mathematical conceptions: Reflections on process and objects on different sides of the same coin. *Educational Studies in Mathematics, 22*(1), 1–36.

Steen, L. A. (1988). The science of patterns. *Science, 240*(4852), 611–616.

Tabor, P., Dibley, D., Hackenberg, A., & Norton, A. (2020). *Numeracy for all learners: Teaching mathematics to students with special needs.* London: Sage Publishers.

Ulrich, C., & Norton, A. (2019). Discerning a progression in conceptions of magnitude during children's construction of number. In *Constructing number: Merging perspectives from psychology and mathematics education.* New York: Springer.

van der Waerden, B. L. (1935). Nachruf auf Emmy Noether. *Mathematische Annalen, 111*, 469–476.

Wigner, E. P. (1960). The unreasonable effectiveness of mathematics in the natural sciences. *Communications on Pure and Applied Mathematics, 13*, 1–14.

Index